Birdscaping *for* Wisconsin *and the* Great Lakes Region

Birdscaping *for* Wisconsin *and the* Great Lakes Region

The 50 Best Native Plants to Attract Birds to Your Midwestern Garden

Mariette Nowak

THE UNIVERSITY OF WISCONSIN PRESS

publication supported by a grant from
The Community Foundation for Greater New Haven
as part of the Urban Haven Project

The University of Wisconsin Press
728 State Street, Suite 443
Madison, Wisconsin 53706
uwpress.wisc.edu

Printed in the United States of America

A Library of Congress Cataloging-in-Publication record is available at https://lccn.loc.gov/2024052554.

ISBN 9780299353049 (paperback)

For my family, especially my three grandsons.

May they always wake to the song of birds,
walk among wildflowers, and thrive,
as one species among many,
in a world of wonders.

Contents

Acknowledgments

The many photographs in this book illustrate the outstanding beauty of birds and the native plants with which they prosper. They've given life and color to my words.

I'd like to thank all the photographers who shared their photos with me: Ryan Brady, Susan Damon, Wendy Gochenaur, Karen Johnson, Margaret Knight, Carolyn Larkin, Sheila McMahon, Anne Moretti, Mike Reese, Lawrence W. Scheunemann, Dawn-Marie Staccia, and Jean Weedman. Katy Chayka and Peter Dziuk at Minnesota Wildflowers also shared photos from their website. Special thanks are due to Jack Bartholmai and Jeremy Meyer, both phenomenal photographers, who provided the majority of bird photos.

I also am grateful to Jennifer Lazewski and Douglas Tallamy, who, at the request of the University of Wisconsin Press, took time to carefully review my draft manuscript and provide many insightful recommendations. Without a doubt, this is a far better book as a result.

I also deeply appreciate the dedication of all the unsung researchers who labor counting hard-to-see caterpillars, crawling through prairies to record species, surveying unseen birds, and myriad other work, at all hours of day and night in good weather and bad. Without them, we would never know all the intricacies, beauty, and diversity of nature.

My sincere thanks, also, to the gifted staff at the University of Wisconsin Press. My editor, Sheila McMahon, worked with great patience and humor to help with tech issues and fine-tune my manuscript. Art director Jennifer Conn and Distillery Marketing and Design deserve praise for the beautiful cover. I'm also very pleased with the attractive interior design and typesetting work by Brenda and Neil West of BNTypographics West. Lisa Himes has produced an extensive index sure to be of great use to readers. And Alison Shay, publicity manager, continues to work to promote my book, helping to spread the word on the value of birdscaping.

As you can see, it took a village to produce this book—a village of photographers, naturalists, editors and designers, reviewers, researchers, botanists, and birders. I am profoundly grateful to each and every one.

And a special thanks goes to my husband, Dave, who reviewed and edited every page, sometimes several times. And more importantly, he was a wonderful companion through the process and has been throughout our life together.

Birdscaping *for* Wisconsin *and the* Great Lakes Region

A female Ruby-throated Hummingbird nectars on a cardinal flower.

PHOTO BY JACK BARTHOLMAI

Introduction

This book features fifty of the top native plants for birds in the Midwest and Great Lakes Region, extending to the Northeastern states as well. The plants selected are those that best attract and support such beloved species as Ruby-throated Hummingbirds, American Goldfinches, Eastern Bluebirds, Indigo Buntings, and our beautiful Wood Warblers.

Planting for birds is the very best way to support birds, especially a wide variety of birds. Bird feeders offer valuable nutrition for birds, especially in winter, but only a small fraction of bird species visit feeders. Moreover, even feeder birds get only a relatively small portion of their food from feeders, especially during the critical nesting period. It's essential to go beyond the bird feeder and plant a variety of native plants that will offer habitat and food sources throughout the year. The recommended plants in this book will provide berries, nectar, nuts, and seeds for birds through the seasons; they will host a variety of protein-packed insects for nestlings and insectivorous birds, and they will provide cover and nesting sites for birds.

By planting native species, you will be replicating the plants with which birds thrived over millions of years. We've already lost three billion birds in the past fifty years in North America. Alarmingly, seventy species are "tipping point species" that have lost half or more of their breeding population and are on track to lose another half in the next fifty years. It's critical to do all we can to stave off more losses. No matter the size of your patch of earth, even if it's potted plants on the patio, you can do your part to keep birds flying, singing, and gracing our lives.

A Magnolia Warbler scans for insects in a flowering crabapple tree.
PHOTO BY JEREMY MEYER, jeremymeyer.photography

The fifty plants in this volume include ten of the top plants for each of five major groups of land birds: hummingbirds, prairie birds, savanna birds, shrubland birds, and woodland birds. In some cases, a single species is featured; in other cases, several species within the same genus are covered. Most are considered keystone species critically important for the survival of birds and other wildlife. All the selected plants can be found at native plant nurseries, although a few are not widely available. Chapter 1 covers the basics of birdscaping—landscaping for birds. Chapter 2 discusses site preparation and planting tips. Chapter 3 shows how to safeguard the songbirds visiting your birdscape. Chapters 4 through 8 feature ten of the top native species to plant to create healthy habitat for each of the major groups of land birds. An appendix offers selections of the top wetland and rain garden plants for birds.

This book is not intended to be a guide to plant identification—many such excellent field guides are available—although occasionally some simple ID tips are provided. Both the common and latest scientific names of the plants are shown. Many scientific names have been changed in recent years due to new DNA analyses by botanists. For example, formerly most American asters were in the genus *Aster* but have been reclassified as *Symphyotrichum*, *Doelingeria*, *Eurybia*, or *Oligoneuron*. For this reason, several scientific names are sometimes given, along with the new ones. The Biota of North America Program was used to document the states in which the plants are native and is widely considered to be the best source for this information. The common names of birds are used in this book; those of non-native species, such as *European Starling*, are italicized. Research by entomologist Douglas W. Tallamy and his research team at the University of Delaware are the primary source for the numbers of caterpillar species supported by various plant species.

An American Robin feasts on berries.

PHOTO BY JACK BARTHOLMAI

1 The Basics of Birdscaping

AN ANCIENT COLLABORATION

Each fall, migrating robins and thrushes fatten up on newly ripe berries, dispersing the seeds en route to wintering grounds. Cedar Waxwings gorge on the berries of eastern red cedar and spread their seeds far and wide, reforesting cedar glades. Hummingbirds sip on the nectar of cardinal flowers and, in doing so, carry their pollen from flower to flower, aiding in pollination and subsequent seed production of these dazzling red flowers. Each spring, migrating warblers feast on emerging leaf-eating caterpillars, thus helping to keep insects under control until the leaves build up their defenses. Such complex interrelationships between birds and native plants have been perfected over thousands of years of coevolution.

The very names of birds, such as the Cedar Waxwing, often indicate the plants with which they've developed intimate associations. The Spruce Grouse thrives on the needles and buds of spruce trees. The Pinyon Jay devours pinyon pine seeds. The Acorn Woodpecker hoards acorns, and the Pine Grosbeak favors seeds of the northern pines. Three-quarters of the diet of the Sage Grouse consists of the leaves and flower clusters of various species of sage.

Other names refer to the plants among which the birds court and nest and spend their lives—the Willow Flycatcher and Willow Ptarmigan, the Cactus Wren, the Alder Flycatcher, the Sedge Wren, the Sage Sparrow, and Sage Thrasher. Then there are names associated with habitat, such as birds of sunny open spaces—Prairie Falcons, Prairie Chickens, and Meadowlarks. And those of woodlands, like the Wood Duck, Wood Pewee, and Wood Thrush, and those in the northernmost forests, like the Boreal Chickadee. The Mangrove Cuckoo, similarly, inhabits coastal mangroves, and the Swamp Sparrow inhabits lowland swamps. All such names suggest the long and commingled evolution of birds and plants.

This ancient collaboration between plants and birds gives us the basic tenet of birdscaping: *Native plants are best for our native birds.* Through millions of years, birds have prospered with the native plants that provided them with the right foods and habitats throughout their life cycles.

Numerous studies are now showing this to be true. In Washington, DC, for example, researchers found that chickadees were unable to raise young successfully on properties with mostly non-native plants, which didn't provide the insects they needed for their nestlings. The insects were unable to survive on these foreign plants because they didn't have the digestive enzymes to feed on these plants with which they hadn't evolved.

Chickadees are among the 96 percent of North American land birds that depend on insects, especially caterpillars, to feed their young. In the study, chickadees often didn't nest at all, and if they did, they had fewer nestlings and a third of the nestlings were less likely to survive. The chickadees were, however, able to nest successfully on similar-sized properties with 70 percent or more native plants. Another study investigated the abundance of the insects in

Chickadees can nest successfully only in landscapes that have at least 70 percent native plants.
PHOTO BY JACK BARTHOLMAI

hedgerows. Hedgerows with invasive plants, like bush honeysuckle, multiflora rose, and Callery pear, had 91 percent fewer caterpillars than hedgerows with native plants. With these studies in mind, aim for at least 70 percent native plants in your yard to ensure the success of your favorite birds.

Use Regionally Native Plants

It's important to use regionally native plants, not plants native to other countries or to states far from your own. Regional or local native plant nurseries will usually carry locally sourced seeds and plants, often called local ecotype or genotype. Plants from sources like these will likely thrive best in your garden or larger landscape and are best adapted for local native birds and wildlife.

Right Plant, Right Place

As with all gardening, it is essential to plant the right plant in the right place. All plants have unique requirements, although many grow in a fairly wide range of conditions. For the best success, match the conditions preferred by the plant of your choice to those of your site. Does the plant like moist or dry soils? Is it best adapted to sand, loam, or clay soils? Sun, partial sun, or shade? Specific requirements are provided for all the plants recommended in this book.

Plant a Diversity of Species

A diversity of plants will provide the greatest benefit to birds. Research has shown that the more species in an area, the more stable and resilient it is. For example, years with greater precipitation will likely favor some species more than others and vice versa for drought years, but the plant community as a whole prospers. The plants suggested in the chapters ahead can serve as the beginnings and basis of your birdscaping efforts, but consider adding more species in the years ahead. Native Plants Database at the National Audubon Society's website provides a list of plants best for birds in your area. Native Plant Finder at the National Wildlife Federation's website helps you find the best native plants for your area that attract butterflies and moths and the birds that feed on their caterpillars, based on the scientific research of Douglas Tallamy. In both cases, simply enter your zip code to obtain a personalized list for your area.

A landscape with a diversity of species offers a variety of insects and seeds for birds and is more resilient to climate change.

PHOTO BY MARIETTE NOWAK

Avoid Cultivars and Treated Plants

In selecting additional plants to create more diversity, stick to straight native species rather than nativars, which are cultivars of native plants. Due to the traits selected in producing these nativars, they often (although not always) lose other traits that are invaluable for wildlife. They may produce less or no nectar, pollen, seeds, fruits. Or they may be less edible, or even nonedible, for the insects needed to nourish birds. Since it is impossible for researchers to study all our many native plants, it is usually best to stick with straight natives. Nativars can usually be identified by checking for a special descriptive name in addition to their species name. For example, "Ginger Wine Ninebark, *Physocarpus opulifolius*" is a nativar, whereas the straight native species would be labeled simply "Ninebark, *Physocarpus opulifolius*." Buy straight for best results.

Also, be sure to avoid any plants or seeds treated with neonicotinoids (also called neonics), which are now the most widely used insecticides on earth. An American Bird Conservancy study in 2013 found that they are extremely lethal to birds. A single seed coated with a neonic can kill a small songbird, like a field sparrow. Lesser amounts weaken birds, harm their reproduction, and disrupt their migration. They also are deadly to bees and many other insects. Neonics are systemic, that is, they are absorbed by the seed and plant tissues. Much of the insecticide comes off in the soil and water, creating long-lasting toxicity, as has been reported in the Midwest and California and other states. Ask your local garden nursery if its stock has been treated with neonics. Whenever possible, buy organic plants and seeds.

BASIC BIRDSCAPING FEATURES FOR EVERY GARDEN

Food and Habitat Through the Seasons

The recommended plants in the chapters ahead will provide both food and habitat, but when possible add more plant species for greater diversity.

A Dark-eyed Junco searches for insects in the leaf litter.
PHOTO BY JEREMY MEYER, jeremymeyer.photography

Leaf Litter

Insects require leaf litter, especially in winter, where they spend the season hidden as adults, cocoons or pupa, or eggs. They, in turn, provide essential food for ground-feeding birds, like towhees, thrashers, thrushes, and some sparrows. In addition, birds use the litter for nest material. Leaf litter also helps retain water for plants and keep the carbon, contained in the roots of plants, stored (sequestered) underground.

Area with Sand or Gravel

Birds have no teeth and will ingest sand and gravel to help grind up the food in their gizzards. Northern Flickers and other birds often dust-bathe, which may help them remove parasites.

Dead Branches and Trees, Fallen or Standing

Dead branches and trees provide insect food, cavities, nest material, and perches for birds. Standing dead trees, called snags, are used as perches by eagles, ospreys, hawks, and owls to scan for prey. Songbirds, like bluebirds, hummingbirds, and others, use dead branches both to search for insects and to rest and roost. Woodpeckers of all kinds, as their name suggests, drill into cavities in dead trees for nesting sites. And a host of other birds, including owls, nuthatches, wrens, bluebirds, chickadees, and wood ducks, use their old cavities or natural cavities in the dead trees for their own nesting. Dead wood also hosts a menagerie of beneficial insects that are prime high-protein food for birds.

Brush, Rock, and Log Piles

Brush piles will attract juncos, sparrows, wrens, and other birds that love the protective cover. I was especially thrilled one April to discover a Winter Wren moving mouselike, as it does, near my brush pile where it had found a haven during its migration north.

Brush piles are easy to construct—simply add any invasive woody plants that you've cut down, branches from pruning your shrubs and trees, or your old Christmas tree and wreaths (if not artificial). There are detailed instructions on building brush piles, but your birds won't be particular. Simply throw

Brush piles provide important cover for birds. The Virginia creeper and grape vines sprawling on this pile at the author's home will soon offer berries as well.

PHOTO BY MARIETTE NOWAK

together a pile of branches, stumps, logs, and brush of any kind. Rock and log piles are more likely to attract rabbits and chipmunks, but these critters will offer additional ready food for owls and hawks that may visit your yard. For an aesthetic touch, plant a vine, like Virginia creeper or a grape vine, to sprawl over your pile and entice more birds with their fruit.

Standing Dead Vegetation Through Winter and Early Spring

In winter, this spent vegetation provides insulating ground cover and seeds for sparrows and finches. Old stalks and seedpods also harbor insects on which many birds feed. Not only that, but the standing stems look stunning when topped with snow.

In spring, resist cleaning up too soon, to avoid killing insects that survived the winter and will attract hungry migrants to your yard. Last year's vegetation and leaf litter will protect both plants and insects against late-season frosts. While you may see some early bees and other insects out and about in early spring, they still need protection during cold nights and thunder storms. The Xerces Society for Invertebrate Conservation gives several suggestions on timing your spring cleanup. If it's safe to plant tomatoes in your area—usually when it's reliably fifty degrees at night—it's probably safe for pollinators and other insects too. Also, if your lawn has grown enough to need a mowing, most insects will not be harmed. In northern states, like the Upper Midwest, it's best to wait until mid to late April. In fact, the longer you can wait the better. Consider joining the No-Mow May (or No-Mow April, depending on your region) movement, which has become very popular in many communities. It's known to benefit bees, butterflies, grasshoppers, and many other insects that thrive on the dandelions, clover, and violets on unmown lawns, which often bloom earlier

than other flowers. And your birds will be better nourished as well, with a greater number and diversity of insects available as they begin to nest.

Water

All birds need water for drinking and bathing, and some need water bodies for breeding and feeding. Birdbaths, ponds, and streams will attract a wide variety of birds. Moving water, in particular, is an irresistible magnet for birds. Consider adding a dripper or mister to birdbaths. Chickadees, finches, and titmice will often land on dripper sprouts to sip from the end of the spout. Ponds or streams with shallow waterfalls also entice many birds, some of which don't come to feeders. In summer, we have enjoyed gorgeous Scarlet Tanagers come to bathe in the shallows below a little waterfall in our yard, a real treat since tanagers don't come to our feeders and are otherwise seldom seen, as they prefer nesting high in leafy trees.

DESIGNING YOUR BIRDSCAPE

Thoughtful design can make your birdscape appeal as much to people as to birds. Following are a few tips to consider as you plan. Also check out online native garden designs, such as those provided at the Wild Ones website. The site features a growing number of free, downloadable native garden designs created by professional landscape designers for multiple ecoregions in the United States, taking into account various light, soil, and moisture conditions.

Layout

Use asymmetrical shapes for your planting bed. Graceful, undulating lines mimic the meandering paths of rivers and the borders between various plant communities in nature. Nonetheless, it's perfectly fine to opt for a more formal layout, especially in the front yard if you are concerned about neighbors.

A Scarlet Tanager bathes in the shallows above a small waterfall at Sylvan Cascades, a water garden at the home of Anne and Patrick Moretti in Dousman, Wisconsin.

PHOTO BY ANNE MORETTI

The gracefully curving lines of this front yard native flower bed at the home of Karen Johnson and Jim Toft in Milwaukee, Wisconsin, mimics the meanders found in nature.

PHOTO BY KAREN JOHNSON

Views and Focal Points

Retain beautiful views, such as the setting or rising sun, a picturesque tree or shrub, or a pond or stream on or adjacent to your property. Consider, as well, views from windows, porches, and benches.

Group Plants Together

For masses of color, plant numbers of the same species in clumps or drifts. This adds attractive visual impact and also provides a better visual signal to birds and pollinators. An odd number of plants, with a minimum of three of each species, gives a more natural look and ensures better cross-pollination.

Variety of Textures

Use different textures to add interest and depth to your garden: plants of different sizes and shapes, plants with coarse and fine stems and branches, plants with smooth or rough bark, plants with a variety of leaves—lobed and unlobed, simple and compound.

All-Season Beauty and Bounty

Choose plants that will provide not only color and interest through the seasons but also yearlong food and habitat for birds. The recommended plants in this book were selected for those purposes, but keep them in mind as you expand your birdscape.

Frame Your Birdscape with Edging

Possibilities for edging include a mowed edge, a rustic fence, or a rock border. Edging accents your planting and will likely please neighbors as well.

The guidelines in this chapter will help you create a beautiful native garden right outside your doorstop.

PHOTO BY MARIETTE NOWAK

2 Site Preparation and Planting

SITE PREPARATION

Locate Underground Utility Lines

It's essential to ask your local utility to flag these lines before you begin any landscaping work for your own safety and to avoid outages or disruptions of utility services.

Remove Invasive Plants

Invasive plants are species that grow outside their historic native range. No longer kept in check by the diseases and animals in their former home ranges, they spread aggressively and outcompete native species. Some of the more common invasive plants in the Midwest include shrubs like buckthorn and multiflora rose, as well as herbaceous species like garlic mustard and dame's rocket. Invasive species will seriously compromise your landscaping efforts and should be eliminated before you start planting native species. There are a variety of best methods that have been found to control different invasive species, such as the careful application of controlled burns, herbicides, biocontrol, grazing, mowing, or hand pulling. Check with your state's department of natural resources or similar organization to learn the best methods to remove specific invasive species. An excellent source of information for the Midwest is the online Wisconsin Manual of Control Recommendations for Ecologically Invasive Plants, which gives detailed coverage of more than twenty-five invasive species.

Prepare Your Planting Bed

It is critical to remove all weeds and other vegetation before you begin to plant. Only a few aggressive rhizomatous weeds, like quack grass and Canada goldenrod, can quickly regain dominance and outcompete your newly installed native plants. The mantra for site preparation is simple, says Neil Diboll of Prairie Nursery in Wisconsin: "Take No Prisoners." Common methods of removing vegetation include smothering, sodcutting, tilling, and applying herbicides.

Smothering

Smothering involves covering a site after it has been mowed as short as possible with material to kill underlying vegetation. Cardboard along with newspaper is

Arborist wood chips are ideal for smothering weeds and building soil.
PHOTO BY MARIETTE NOWAK

often recommended and considered to be a sustainable way to recycle material that might otherwise go to landfills. But new research has found that corrugated cardboard contains high levels of PFAS chemicals, the "forever chemicals" that never degrade and contaminate both soil and water.

As an alternative to cardboard and newspaper, Linda Chalker-Scott, a professor emerita at Washington State University, recommends using arborist wood chips—chipped branches and leaves fresh from arborist crews—not bark mulches. Her research has found that arborist wood chips not only are ideal for weed control but also "build soil, reduce evaporation, prevent soil erosion and compaction, and enhance beneficial microbes, especially mycorrhizae." A thick layer of at least six inches is best for weed control, and more may need to be added to maintain a four-inch layer.

Another excellent mulch for smothering weeds is leaf litter covered with natural burlap, a method popular among some organic gardeners. Use a three-inch layer of shredded leaves; avoid whole leaves, which will prevent rain from penetrating the soil. If autumn leaves aren't available, weed-free straw can be used as an alternative. Cover the leaves or straw with the natural burlap, securing it with landscape pins. The burlap, which is an organic, breathable product, will break down after a few seasons and can then be composted. Whichever of the smothering methods you use, keep it in place for a full growing season.

Smother crops are recommended for multiacre sites. They are annual crops, like buckwheat (*Fagopyrum esculentum*) and winter wheat (*Triticum aestivum*), which kill weeds and other vegetation by shading them; they also add valuable organic matter to the soil. Smother crops require two years of use to effectively control weeds.

Sod Cutting

Sod cutting is the most straightforward approach to eliminate a lawn. Remove the top three inches of sod by hand or with the use of a sod cutter, which can be rented from hardware stores. But hauling away the heavy rolls of sod can be backbreaking work, as native plant gardeners Pat and Carl Brust of Wisconsin found. The ground surface also ends up several inches lower than previously. Adding purchased topsoil is not recommended, as it often contains weed seeds. Another drawback to sod removal is that some weeds with deep roots remain intact, as native landscapers Susan and Paul Damon in Minnesota discovered when dandelions showed up later in their planting beds.

Tilling

Tilling requires lifting and turning soil every three weeks until there is no regrowth and through a full growing season, especially if weeds like quack grass and Johnson grass are present. Both have rhizomes, horizontal stems below the ground surface. When broken up during tilling, their stems break into many small pieces, all capable of producing new plants and making eradication extremely difficult. There are significant environmental costs to tilling, such as air pollution from the rototiller's fuel, although this can be minimized with the newer electric rototillers. In either case, heavy equipment can compact the soil.

Another major disadvantage of tilling is that it destroys the soil structure and releases the carbon stored in the soil, thus contributing to climate change. Perhaps worst of all, tilling destroys beneficial soil organisms. The Natural Resources Conservation Service's Soil Quality Team describes tillage as a soil catastrophe: "Tilling the soil is the equivalent of an earthquake, hurricane, tornado, and forest fire occurring simultaneously to the world of soil organisms. Simply stated, tillage is bad for the soil."

Herbicides

For all the reasons given here, many experts favor herbicide treatments over tilling. Site preparation with herbicides has a number of advantages. A one-time application of herbicide is usually the quickest and least expensive way to eradicate sod, although old fields may require an additional application or two. A more important advantage is that the soil structure remains intact and weed seeds in the soil are not brought to the surface to germinate. Also, the carbon contained in the roots of plants (killed along with their aboveground stems) remains underground and is not released into the air, where it would add to the carbon load causing climate change. Additionally, this method prevents soil erosion, retains organic matter that enriches the soil, and saves the labor and possible fuel costs of hauling the sod (or other vegetation) away.

Consider all the pros and cons of the various methods of site preparation. If herbicides seem like the best choice, then use them sparingly and selectively.

PLANTING HERBACEOUS SPECIES

Plants are more expensive than seeds but will give faster results. They are an especially good choice for front yards in urban and suburban areas where they can simply replace traditional non-native perennial beds and borders. Plants also bloom earlier, often in the first season, and can be easily added to existing plantings.

Plants can be purchased bare rooted or in pots. Whichever you buy, follow the instructions given by the nursery where you purchase the plants. Following are a few additional guidelines that may be helpful. Plants are best installed when they are dormant and not in flower in early spring. Early fall with its cooler temperatures and often more rainfall is also a good time and gives plants sufficient time to put down their roots before winter.

Bare Root Plants

Bare root plants are those with fibrous roots or plants in the form of rhizomes, bulbs, or corms. They should never be allowed to dry out; even a minute in the

Native plant sales, sponsored by Wild Ones and other groups, offer buyers a variety of species.
PHOTO BY JEAN WEEDMAN

sun or wind can kill them. Cover them with damp burlap while digging a hole deep enough to accommodate the entire length of the roots. Fibrous plants should be set so that the buds just above the roots are about one inch below the ground surface. Plants with rhizomes should be placed horizontally about an inch or two deep with the buds at ground level. Corms should be planted so that their buds at the top are about two inches below ground level. Bulbs need to be positioned so that any green part, indicative of aboveground growth, is above the soil surface. Make sure there is good root contact with the soil, as you fill the hole. Then press the soil surface to eliminate air bubbles, lightly mulch, and water thoroughly.

Potted Plants

Water the potted plants thoroughly before installation; dry root balls will repel water, even if the surrounding soil is wet. Dig a hole about the same depth as the pot. After removing the plant with the surrounding soil from the pot, check the roots and loosen them. If tightly interwoven, cut off a thin slice from the bottom of the root ball and make a cut halfway up the crown to stimulate growth. For plants in four-inch or larger pots with extremely matted roots, cut the root ball two-thirds of the way up the plant in several places. Place the plant in the hole so that its soil surface is slightly below that of the surrounding ground. This ensures that rainwater will run toward the plant, not away from it. Add more soil if needed and then tamp down the soil around the plant to prevent air pockets that can dry it out. Cover the soil with a light mulch and water thoroughly.

PLANTING SEEDS

Seeds take longer to establish, generally about three years, but are far less expensive. They can provide a more diverse mix of species since many more native species are available as seeds than as plants. They are also the most cost-effective way to plant large areas. For information on the propagation of particular plants, see William Cullina's *The New England Wildflower Society Guide to Growing and Propagating Wildflowers of the United States and Canada.*

As with herbaceous plants, spring and fall are the best times to install seeds. After seeding, mulch with a light application of weed-free straw or compost. In winter, "frost" seeding can be done over a thin layer of soft snow; the seeds will work themselves into the soil gradually during the freeze/thaw cycles.

Seeding Former Lawns and Other Small Sites

Smothering for a full growing season or applying an herbicide in fall are generally the best methods to remove vegetation before seeding a lawn or relatively small site an acre or less in size. In spring of the following year, lightly till the soil surface to prepare a good planting bed. Often a cover crop like native Canada Wild Rye (*Elymus canadensis*) is added with the other native seeds. This grass begins growth in the early cooler season, helps reduce weeds and erosion, and provides shade for slower-growing species; later it disappears, as other prairie species mature. It's also a good idea to add sawdust to the seed mix to help spread the seed. After seeding, lightly rake in the seed and walk over the area or use a roller to ensure good seed contact with the soil. Cover with a light layer of weed-free straw and water regularly for a month if possible.

Seeding Multi-Acre Sites

Smother crops, tilling, or the use of an herbicide are the primary means of removing vegetation from large sites. Because tilling, as mentioned, causes great environmental damage and soil destruction, many experts prefer smother crops or herbicides. Smother crops usually require two years of use to effectively control weeds. Herbicides, used judiciously, are also effective. Seeding can be done after the weeds have been killed. The use of a no-till seeder is best because it minimizes soil disturbance and doesn't expose underground weed seeds to sunlight where they will germinate. As in smaller sites, cover crop seeds are often used in the seed mix. After seeding, the area should be rolled to ensure good seed contact with the soil. And, as with smaller sites, a light layer of mulch is recommended. Much more advice is available from native nurseries and their online websites.

PLANTING WOODY PLANTS

The best time to plant trees and shrubs is during the early spring before growth begins and before the heat and drought summer may bring. Fall is the next best time, after the woody plants have lost their leaves. The plants can develop their roots before winter's freezing temperatures.

The spacing of plants is somewhat arbitrary. Allow sufficient space for specimen plants, like pagoda dogwood, to display their full stature and beauty. Generally, smaller shrubs are spaced about two to three feet apart, large shrubs about four to six feet apart, and trees about ten to thirteen feet apart. But some experts are now suggesting that we mimic the spacing found in natural woodlands, where trees are often much closer together.

Bare Root Plants

These are often available in large quantities from state agencies and local nurseries. As with bare root herbaceous plants, it is essential to keep the roots moist until planted. To create a hole for small plants, step on a spade, pushing it full length, and force it forward to create an opening. Insert the plant's roots and press the soil back in place to ensure good soil contact and eliminate air pockets that could dry out the root. For larger plants, form a mound of soil in a hole deep enough for the roots and spread them over the mound, cover with soil, and tamp firmly.

Balled and Burlapped and Potted Plants

Dig a hole a little more shallow than the root ball and two times as wide. A wide hole will allow the roots in the root ball to expand throughout the top foot of the soil. Loosen the soil on the sides and bottom of the hole so roots can penetrate more easily. Remove any string, wire, plastic wrapping, or burlap surrounding the root ball. Examine the roots as quickly as possible to avoid drying them. If the roots encircle the root ball, slice through the ball on several sides.

Rotate the plant in the direction you prefer and place in the hole, spreading the roots downward and outward. Plant so the root flare (where the trunk starts to bulge out at the bottom) is slightly above the soil surface, about one inch higher (this is in contrast to the rule for herbaceous species, which are typically planted level with the soil in which they are grown). Planting too deep is a major cause of problems for trees and woody plants.

Fill the hole with soil gradually, settling with water as you go. Use any remaining soil to build a small berm around the edge of the planting hole to retain water.

Acorns and Seeds

There is evidence that planting seeds and acorns directly is an effective way to plant shrublands and woodlands, often more successful than transplanting plants germinated on other sites. For more information on acorn planting, see Planting an Oak in chapter 6. For information on propagating other woody species from seed, see William Cullina's *Native Trees, Shrubs, and Vines.*

Cuttings from Wetland Trees and Shrubs

Some wetland woody plants can be grown from stem cuttings that are simply placed in moist soil where they will take root on their own. The stems should be cut when the plants are dormant in early spring or late fall or winter when the ground is not frozen. Stems should be two to three feet long and about one-half inch to two and one-half inches wide. Cut the lower ends at a slant and the top end flat so they can be pounded into the soil. After cutting, place the stems in water and plant the same day or within a few days. Pound a rod into the soil to make a hole for the cutting, then place a block over the flat end of the stem to pound it into the hole to avoid splitting the stem.

Wet-loving willows and dogwoods root easily from cuttings. Some other shrubs recommended in this book that root fairly well from cuttings include prairie willow, common elderberry, and nannyberry viburnum.

CARE AFTER PLANTING

Fertilizing

The use of fertilizers is not generally recommended for native plants, which have grown successfully for thousands of years without them. Fertilizers tend to give the edge to invasive plants, which can aggressively compete with the natives.

Watering

After installing, plants and seeds need to be watered regularly. If possible, keep seeds moist until they germinate and begin to establish. Transplants should be watered deeply about once a week when first installed, then every other week through the first season to encourage deep roots. Avoid shallow watering, which encourages shallow roots and makes plants less adaptable when watering stops. Most plants need about one inch of water per week. Sometimes woody plants will need watering through the second growing season. Watering during droughts is also essential, especially for woody plants that are not easily replaced. Water slowly and wisely, using timed meters, soaker hoses, and other means to conserve this vital resource. Smaller plants and smaller areas will need shorter watering times compared with larger areas and larger plants, like trees.

Mulching

Always mulch after installing plants. Mulch will suppress weeds, help keep the soil moist, prevent erosion, and moderate soil temperatures. Woody plants can receive up to three inches of mulch, while herbaceous plants should receive no more than one inch. The larger the mulched area around woody plants the better; a circle of three feet in diameter is common. Always keep mulch away from contact with the stems to avoid stem rot.

Some of the best mulches are those recommended for smothering unwanted vegetation– fresh wood chips and shredded leaves covered by natural burlap. Weed-free straw is also a good option. But use thinner layers of these mulches, as recommended in the section on smothering. For slopes, keep mulch in place with an erosion-control mat.

Protection from Herbivores

Herbivores, such as mice, rabbits, and deer, can damage many new plantings. Fencing, sprays, tree shelters, and wraps can deter them.

Deer are, by far, the greatest threat to birdscapes, whether newly planted or well established. They chomp on wildflowers and browse on woody plants. In addition, they rub their antlers on tender-barked trees from early fall to late winter, often killing them. Vegetation lost to deer can result in devastating declines in bird populations.

Repellents, fencing, and planting species less favored by deer can stop or reduce deer damage. Repellents made from rotten eggs, either homemade or purchased, are unpalatable to deer but must be applied after every rain. Fencing is a more permanent solution but needs to be six to seven feet high. Electric fencing is the most reliable: install a strand about four feet high, baited every few yards with peanut butter or salt to attract deer that get a shock when they try to eat the bait. Many experts suggest planting species less favored by deer, but I have looked at lists of such plants and found a great deal of conflicting advice. If hungry and overpopulated in an area, deer will eat anything. The most effective long-term solution for controlling deer populations is through hunting or sharpshooting, so encourage these in your communities whenever possible.

Rabbits and mice also can greatly damage new plantings. Rabbits are a threat especially in spring and early summer, devouring tender new seedlings and leafy greens. The best defense is a three-foot wire-mesh fence around new plantings. Smelly substances like those used to deter deer are also helpful.

Mice and voles, as well as rabbits, can be a problem in winter when they gnaw on the bark at the base of trees and shrubs. They often do so under the cover of snow or mulch, so keep mulch three to four inches away from the base of woody plants. Other good options are tree shelters or material wrapped around the trunks of young woody plants for a year or two until they are less susceptible to damage.

Weeding

Getting rid of weeds is vital during the first few years, while plants are getting established. Doing so can pay big dividends later. Weeds are always easier to deal with when small. But don't waste time removing annual weeds that are not invasive and will die out. The trick is being able to identify weeds when small. A good source for the identification of fifty-four weed species is *Common Weed Seedlings of the North Central States*, a Michigan State University publication. To ensure you identify and retain the desirable native plant seedlings, see *The Tallgrass Prairie Center Guide to Seed and Seedling Identification in the Upper Midwest* by Dave Williams. Also try to get help from local Wild Ones chapters, native plant societies, or nature centers. Another technique is to plant a few seeds of each native species in a pot to help distinguish "invited" seedlings from unwelcome ones.

In the first year, scout for weeds every two weeks; this can be done less often in later years. If weeds are numerous in new prairies, mow them to about six inches with a weed whip or flail mower. Native prairie plants seldom get more than six inches during the first year, so they will not be affected.

Prescribed burns are used to manage established native prairies and woodlands, but an unburned portion should always be provided as a refuge for insects.

PHOTO BY MARIETTE NOWAK

Long-Term Maintenance

Prescribed burns or mowing are recommended to maintain established prairies and oak/hickory woodlands. Burning removes accumulated thatch that can repress plant growth, exposes the soil to the sun, can help slow the invasion of woody species and weeds in prairies, and stimulates soil microorganisms that improve plant growth. Mowing is a good option in urban/suburban areas where prescribed burns aren't allowed.

Whether mowing or burning, keep the birds in mind. It's best to act in very early spring before birds are nesting. Late fall burning or mowing removes seeds and cover for birds but has been found to benefit some prairie specialists, like Northern Harriers and Upland Sandpipers, that have better nesting success, presumably because more fresh cover develops by nesting season. Burns at any time of year can kill insects, which are invaluable to birds. For this reason, it's wise to burn only a third to a fifth of an area to preserve insects and their eggs and pupae on the unburned portion. The insects can then spread to the entire area during the next growing season.

Prescribed burns can be dangerous and need to be carefully planned and controlled. Usually a permit is required from the local sheriff or other agency. It's best to get professional help if possible, especially if you are inexperienced. Prescribed burn training is also offered by the Nature Conservancy and other restoration organizations.

The Common Yellowthroat often nests in wet thickets.

PHOTO BY JEREMY MEYER, jeremymeyer.photography

3 Safeguarding your Songbirds

The native plants in your garden are sure to attract a variety of birds. But for your birds to thrive, you will also need to protect them from a number of common hazards, including cats, windows, pesticides, and poisons. Many billions of birds die each year across the country from these hazards.

PROTECT BIRDS FROM PREDATION BY CATS

My family and I have always loved cats—handsome, cuddly, and fascinating—and we've enjoyed them as pets for years. When our sons were young, each had a cat of his own. At first, we unknowingly allowed the cats to roam outdoors at will. They occasionally showed up at our door with their hunting trophies, mostly shrews. But one day our son Rob saw our cat Tiger kill a robin! We were shocked to find that our cats were killing the very birds that we were trying to attract to our yard and immediately began to keep them indoors. We sometimes put our cats on leashes outside our door, where they were often harassed vocally by grackles. We also enticed them to enjoy the indoors with scratching posts, toys, and play time.

The domestic cat, *Felix catus*, is not native to our country nor to most of the other countries around the world where it has been introduced. Ever more information is confirming that cats have wreaked havoc around the globe, killing billions of birds, mammals, and other wildlife wherever they roam. The International Union for Conservation of Nature has listed the domestic cat as one of the one hundred worst non-native invasive species in the world and recorded its role in the extinction of thirty-three birds, mammals, and reptiles.

Domestic cats kill billions of birds and other wildlife each year in the United States.

PHOTO BY MARIETTE NOWAK

In 2013, a team of researchers compiled the findings of twenty-one of the best studies of cat predation and published their findings in the journal *Nature Communications*. They found that cats are likely the single greatest source of human-caused mortality for birds and mammals in the United States, killing a whopping 2.4 billion birds each year, as well as an even greater number, of mammals—23 billion. These numbers confirm that domestic cats, now numbering over 100 million in our country, are likely the largest threat to birds, far surpassing other grave threats to birds, such as window collisions and habitat loss.

Cats should always be kept indoors. Even well-fed cats are lethal predators; their hunting instinct is innate and independent of hunger. The lives of countless birds and other native animals would be saved if everyone did so.

Not only is keeping cats indoors better for birds; it's also better for the cats themselves. Free-roaming cats can be killed by cars, attacked by dogs and coyotes, poisoned, stolen, or abused. They can also contract fatal diseases like rabies, viruses, or feline distemper. Outdoor cats live only two to five years, while indoor cats can live to eighteen years or more. Because of all the hazards faced by outdoor cats, veterinarians and humane societies alike support keeping cats indoors.

Related to these facts, Trap, Neuter, Release (TNR) should be mentioned. TNR is a program by which feral cats are trapped, spayed, or neutered and then basically abandoned back outdoors to face the same hazards mentioned here. TNR, although well intentioned, has been aimed at reducing the number of feral cats, Unfortunately, it has failed miserably. To begin with, it is impossible to neuter and spay enough cats to affect their population numbers. Long-term studies of TNR colonies in Florida and California by a team of researchers concluded that although some of the original cats died off, the colonies grew in size because of the dumping of unwanted cats and the attraction of additional stray cats to the food provisions at the colonies. TNR is strongly opposed by the American Bird Conservancy, the Cornell Lab of Ornithology, and other wildlife experts. Cat populations were not reduced, and the cats remained dangerous predators of birds and wildlife.

Our family, as mentioned, is able to have happy, long-lived indoor cats by providing them with scratching posts, toys, and play time. Often, we've had two cats who enjoyed being playmates.

Many ideas and products are available today, now that more and more cat owners are eager to keep their cats safe and content indoors. Indoor cats enjoy cat trees, perches, and window seats from which they watch outdoor activity. A cat exercise wheel, similar to the famous hamster wheel, is also now available commercially. DIY toys can be made from crumbled paper, cardboard boxes, toilet paper rolls, and socks.

There are also many options to keep pet cats safe outdoors, giving them plenty of fresh air and sunshine. Special cat harnesses and leashes, as well as backpacks and strollers, are made for feline comfort. Cats can also be given their own personal space in your yard with cat-safe fences, The ultimate option is a catio—a permanent or portable (take it on a picnic!) enclosure for cats.

PREVENT BIRD COLLISIONS INTO WINDOWS

Clear glass is invisible and deadly to birds. During the day, glass windows can mirror the sky and surrounding vegetation, and birds will inadvertently try to fly right through, crashing into the glass. Bright night lights in windows and on buildings disorient birds, especially during migration, and often cause them to collide into windows and buildings.

A 2014 study showed that window collisions kill up to a billion birds in the United States each year and have taken a toll on both common and rare birds. Although bird crashes into windows on high rises have gotten the most publicity, nearly half of bird collisions occur at homes one to three stories high. Some of our favorite birds, like Ruby-throated Hummingbirds, Cedar Waxwings, and Black-and-white Warblers, are particularly vulnerable to home windows. Overall, many warblers—some of the most beloved, beautiful, and sought-after birds—experience an extremely high collision risk with ten among the twenty-five most vulnerable species.

The beautiful Black-and-white Warbler is extremely vulnerable to window collisions.
PHOTO BY JACK BARTHOLMAI

Anything that reduces window reflections will save birds' lives, and many effective solutions are now available. Fortunately, there are usually only a few windows in your home that pose the worst threat to birds and you can start with these. Many window treatments are easy to install, but window washers, local handypersons, or house painters may be able to help if you cannot do so yourself.

The American Bird Conservancy has tested many window treatments, and its website offers a wealth of information to help you select one that will suit your circumstances. Old-fashioned screens on the outside of windows, which were once common, offer one of the best solutions. Screens not only break up reflections but also cushion any bird that might fly into them. There are now screens available that can be installed on windows and doors with suction cups or screws.

Warbler Species at High Risk of Window Collisions

This list includes warbler species at particularly high risk from colliding into windows of buildings of all heights, from home residences to high rises.

Black-and-white Warbler
Black-throated Blue Warbler
Canada Warbler
Cape May Warbler
Common Yellowthroat
Connecticut Warbler
Golden-winged Warbler
Kentucky Warbler
Louisiana Waterthrush
Mourning Warbler
Nashville Warbler
Northern Waterthrush
Ovenbird
Worm-eating Warbler

Canada Warbler.
PHOTO BY JACK BARTHOLMAI

These zen wind curtains at the author's home stop birds from hitting windows and don't obstruct views from inside the house.

PHOTO BY MARIETTE NOWAK

The American Bird Conservancy also highly recommends three inexpensive, easy-to-install products that have been proven to prevent bird strikes: Zen wind curtains, Feather Friendly dotted strips, and ABC BirdTape. All can be installed as DIY projects with instructions online. Zen wind curtains, also called Acopian BirdSavers, are long-lasting hanging cords spaced about three to four inches apart. These are my favorite window treatment and have successfully protected birds from flying into ten of my largest windows. I find them quite elegant and aesthetically pleasing—so do my visitors—and I still get excellent views of the birds and landscaping from inside my home. Neighbors installed Zen wind curtains at both their home and mine. Two other inexpensive products that the Conservancy recommends are Feather Friendly and ABC BirdTape. Feather Friendly offers durable white or gray dotted squares applied in strips, while ABC BirdTape is a durable translucent tape in strips or squares. You'll find many more solutions on the Conservancy's website.

Feather Friendly dotted strips can be applied to the outside of windows to make the glass visible to birds.
PHOTO BY CAROLYN J. LARKIN

All window treatments, including those mentioned here, should be applied on the outside of windows. If window reflections are strong, internally applied products are often hidden from outside, making them ineffective in preventing collisions. Some products also need to be replaced regularly to remain viable.

Fortunately, there has been progress on protecting windows on taller buildings, where 56 percent of bird collisions occur. Cities and states from New York to California have recently passed bird-safe building measures. At Chicago's McCormick Place Convention Center, where nearly a thousand birds died on a night in October 2023, bird-safe film has now been installed on the windows, which will help prevent such tragic occurrences. And the city has also mandated the use of bird-safe materials in the construction of new buildings and the renovation of existing buildings.

In Milwaukee, Fiserv Forum is the world's first bird-friendly arena. The windows on the massive building have a thin ceramic coating on glass, called fritting. It gives the glass tiny lines that humans can see up close but does not interfere with the transparency. For birds, the coating reduces the transparency of the glass and appears to them as a wall that they should avoid. Also in Milwaukee, Marquette University's new nursing school has bird-safe glass in strategic locations; the glass contains a small dot pattern designed to steer birds away from large windows.

In New York, legislation was introduced after the death of the beloved Flaco, a free-flying Eurasian Eagle-Owl that had escaped from the Central Park Zoo but later crashed into a building. Called the FLACO (Feathered Lives Also Count) Act, it would require bird-friendly designs on large, new and renovated state buildings.

Support such bird-safe measures wherever you live.

AVOID PESTICIDES AND POISONS

Consider the brilliant Baltimore Oriole arriving in your yard in spring after a hazardous trip of more than a thousand of miles from as far as the northern tip of South America. And then, famished on arrival, it sickens and perhaps dies from eating caterpillars that have ingested poisons sprayed on a cultivated rosebush. Despite their spectacular feats of migration, birds are extremely vulnerable to the many pesticides commonly used in yards and on farms.

In New York, the FLACO Act recalls the beloved owl that died by crashing into a building.

Birds like the Baltimore Oriole can die by eating insects poisoned by pesticides.
PHOTO BY JACK BARTHOLMAI

Songbirds can be harmed by pesticides directly when they feed on seeds treated with pesticides or indirectly when they eat insects that have themselves eaten pesticide-laden food. Pesticide use also causes the direct loss of insects themselves, depriving birds of an essential food source, which can lead to poor nesting success, even starvation. Hawks, owls, and other birds of prey can die after eating rats that have fed on baits laced with poisons.

Pesticides include toxins used to kill insects (insecticides), plants (herbicides), rodents like rats (rodenticides), and mildew or rusts (fungicides). They are used as sprays, coatings, dustings, and baits, and they are deadly—not only to their intended targets but also to many birds, bees, and a variety of wildlife. Up to

72 million birds alone are killed each year by pesticides, according to the Environmental Protection Agency. They have sickened, and sometimes killed, humans as well.

As mentioned in previous chapters, there are times when the judicious use of pesticides may be needed to kill weeds on a site in preparation for landscaping with native plants in order to provide important long-term habitat for birds and the insects they need. One-time applications may also be needed to treat certain pernicious invasive plants, like buckthorn and crown vetch, especially those with deep or fibrous roots—again used sparingly with the intent of providing permanent habitat for birds and wildlife.

In most other situations, however, pesticides should be avoided on lawns and in gardens. In particular, never use neonics. Even the tiniest seeds coated with these pesticides cause great harm, even death, to birds. Neonics impact bird migration, nesting success, and critical body functions. Some experts are now calling neonics the new DDT, relating it to the pesticide that caused the near extinction of eagles and hawks until it was banned in this country in 1972. Other countries, such as the European Union and United Kingdom, have banned or greatly restricted neonics, but the United States has not done so yet. Some states, however, including Maine, New Jersey, Vermont, and California, have taken the lead, enacting bans or restrictions on neonics. Unfortunately many common backyard lawn and garden products contain neonics. I found a number of these products at my local hardware store in the small town of East Troy, Wisconsin. To determine if a pesticide contains neonics, look at the labels. Imidacloprid, acetamiprid, dinotefuran, clothianidin, thialoprid, and thiamethoxam are all neonics.

By avoiding harmful pesticides, gardeners and native plant enthusiasts can play a vital role in preserving insect abundance and diversity, which, in turn, will ensure our birds and their nestlings will thrive. According to the Xerces Society, recent research has shown that pollinators are faring better in urban and suburban environments where they are protected from the widespread use of pesticides so common on farmland.

Instead of using chemicals, consider tolerating minimal damage to foliage.

Even though native plants support insects and you're likely to see some insect damage, you don't usually have to worry that they will defoliate your garden. One study showed that homeowners who planted natives exclusively had only 3 percent leaf damage. The birds and other natural enemies of the insects kept the pests in check. In fact, if your vegetation is perfect, especially later in the growing season, you should be alarmed as it suggests that you have no insects for birds.

When weeding is necessary, try the following nonchemical bird-safe methods of dealing with them.

Weed by Hand

Many weeding tools are available that make weeding easier and more efficient than ever. Some can even be used without bending down. Many people find weeding a way to relax and a time to enjoy close up the diversity of plants and wildlife in their gardens. Some experts now recommend cutting weeds at ground level because it does not disturb the soil nor does it uncover more weed seeds to germinate and make the situation worse. This can seem tedious, but Larry Weaner, a leading native landscape designer, found that it took him the same time to cut four weeds at their base as to pull one by the roots (depending on the weed, of course). The weed will grow back if the area is not mulched and is open to sunlight, but this is less likely if shaded by existing vegetation.

Mulch not only suppresses weeds but makes an attractive border around plantings.
PHOTO BY SHEILA MCMAHON

Using mulch instead of chemicals to suppress weeds will ensure that birds like this Tennessee Warbler can find the insects they need for their own nourishment and for that of their nestlings.

PHOTO BY JACK BARTHOLMAI

Weed Early in the Growing Season

Plants are easiest to weed when plants are small and have less developed roots. These weeds can be thrown into the compost pile to provide rich organic compost for future use.

Mulch to Suppress Weeds

Shredded leaves and arborist wood chips are among the best mulches, as mentioned in chapter 2. These mulches will also help retain moisture in the soil and add valuable organic matter.

Use Corn Gluten to Prevent Weed Germination

Corn gluten is a natural byproduct of corn that prevents seeds from germinating but will not harm established plants. It is often recommended to prevent new weed growth on established lawns.

A hummingbird nectaring on a cardinal flower.
PHOTO BY DAWN-MARIE STACCIA

4 Top 10 Plants for Ruby-Throated Hummingbirds

Hummingbirds flit about flowers like tiny fireworks—flashy, colorful, and quick. They are found only in the Americas. Nowhere else in the world do these winged jewels bedazzle with their unbelievable migrations and acrobatic maneuvers in flight. There are more than 350 species, with about 20 of them summering in the United States and Canada. East of the Mississippi, only one species is common, the beautiful Ruby-throated Hummingbird, although a few of the western species occasionally make an appearance.

Hummingbirds and flowers have coevolved over millions of years, each adapting to mutually benefit one another. Hummingbirds, for their part, evolved bills and tongues suited to feed on flower nectar. Some hummers developed bills perfectly aligned in curve and length with the corolla of a favored flower. And new research shows that hummingbird tongues act like micropumps to draw up nectar much faster than through capillary action, as was once believed.

Plants, in turn, evolved in ways to better entice hummingbirds to visit and pollinate them. Hummingbirds have prodigious appetites and need to devour eight times their weight in fluids each day. Due to their need to consume such large quantities of nectar many times a day, hummers are superior pollinators. Flowers accommodated hummers by stepping up their nectar production. And the flowers benefited because they were more frequently pollinated and produced more seeds. Flowers also evolved shapes to accommodate hummer bills, as the hummers themselves simultaneously evolved suitable bills to fit the flowers.

While nectar constitutes the bulk of the hummingbird diet, small insects make up about 25 percent. These tiny birds feed by darting out from a perch to catch insects in the air or on foliage, and they will grab spiders or trapped insects from spider webs. Hummingbirds are thought to feed their young mostly very small insects, perhaps mixed with nectar. The plants in this chapter offer both nectar and insects for hummingbirds. Eight of these plants have flowers rich in nectar and are especially attractive to hummingbirds. With their varied blooming times, these species will provide nectar for hummers throughout the growing season. Wild columbine and fire pink blossom in succession in spring and summer. Jewelweed is the last nectar-rich flower to bloom in autumn. The other species come into bloom at various times throughout the summer. The other two recommended species are the shrubs, pussy willow and New Jersey tea, which host many small insects fit for the tiny beaks of ruby-throats. Together, all these plants will provide a welcome feast for these winged wonders.

Several of these top hummingbird plants, however, despite being unsurpassed for attracting hummingbirds, can be difficult to grow in some situations. Trumpet creeper, which has the richest nectar of any plant in North America, is nonetheless quite aggressive in ideal conditions. Cardinal flower, a superb favorite of hummers, is often short-lived and can be difficult to maintain. Michigan lily, a gorgeous hummer flower, is fussy about its growing conditions and is often devoured by deer and other wildlife. For these reasons, gardeners may prefer to plant some easier-to-grow alternatives, which are also exceptionally attractive to hummingbirds. The native red honeysuckle vine (*Lonicera dioica*) is a great substitute for trumpet creeper. Wild blue phlox (*Phlox divaricata*),

Michigan lily.
PHOTO BY MARIETTE NOWAK

Hummingbird Flowers

Through their long coevolution, some flowers specialized in attracting hummingbirds by developing a unique set of characteristics to entice these super pollinators. As you check out the flowers in this chapter, look for these characteristics.

1. Hummingbird flowers usually have long tubes with nectar at their bases, easy for the long-billed and long-tongued hummingbirds to tap, but difficult for shorter-tongued insects. These tubes are often thicker at the base to prevent damage from hummers' sharp beaks and to protect from nectar thieves, such as bees who poke holes in the tubes to steal nectar.
2. Hummingbird flowers are often red, orangish-red, or pinkish-red, colors that hummers (and we) see easily, but not bees who see these colors as black. The bright colors also serve as signals to hummers as they zip overhead.
3. Hummingbird flowers often have projecting stamens and pistils. As hummers feast on nectar, pollen on the stamens is brushed on their bodies and later transferred to the pistils of other flowers the hummers visit, enabling pollination to take place.
4. Hummingbird flowers usually lack scent. Hummingbirds locate flowers by sight, not smell, unlike butterflies and bees that are attracted by fragrance. However, it's recently been discovered that hummers are able to smell the chemical scents of some aggressive insects, like ants that may be on flowers, and will then avoid visiting those flowers.
5. Hummingbird flowers often hang downward or to the side. With their amazing agility, hummers can easily turn their heads up or to the side, but most insects can't.
6. Hummingbird flowers lack horizontal landing platforms, which are not needed by hovering hummers but are often required by insects.
7. Hummingbird flowers bloom during the day, unlike night-flowering species that attract moths and bats.
8. Hummingbird flowers provide a large amount of rich nectar with up to 26 percent sugar content.

foxglove beardtongue (*Penstemon digitalis*), and great blue lobelia (*Lobelia siphilitica*) are good alternatives to Michigan lily and cardinal flower. All four are relatively easy to grow in their preferred habitats, widely available and native throughout the Midwest.

WILD COLUMBINE (*Aquilegia canadensis*)

Holly Family, Aquifoliaceae
Perennial Wildflower
Size: 1–3 feet
Color and Season: Red and yellow flowers in spring
Light Requirement: Partial shade to shade
Soil: Dry to moist
Native Habitat: Woods, limestone cliffs
Native to: AR CT DE GA IA IL IN KS KY MA ME MI MN MO NC ND NE NH NJ NY OH OK PA RI SC SD TN TX VA VT WI WV; native but rare in AL FL MS
USDA Hardiness Zones: 3–8

Wild columbine is one of our most exquisite and beloved spring wildflowers. I was thrilled one day to come across a beautiful display of dozens of wild columbine growing in pockets amid a rocky limestone outcrop, their red and

A Ruby-throated Hummingbird sips nectar from a wild columbine.
PHOTO BY LAWRENCE W. SCHEUNEMANN

yellow blossoms like colorful bells dancing in the spring breeze. The name *aquilegia* is from the Latin word *aquil* (eagle) and may refer to the flower's spurs, which are thought to look like the talons of eagles. Alternatively, the name may be based on the Latin *aqua* (water) and *legere* (to collect), in reference to the nectar water collected in the spurs of the flowers. The common name, columbine, comes from *columba* (dove), for the fanciful resemblance of the petals to five doves drinking at a dish.

Wild columbine is the first native wildflower to welcome migrating Ruby-throated Hummingbirds in spring. Ornithologists have found that the hummers seem to time their journeys in sync with the blossoming of the flower, a favorite food source. And no wonder, for the nectar of wild columbine provides both energizing sugars and essential amino acids. Along with its red-colored relatives in the western United States, the red-colored wild columbine native to the eastern United States (*Aquilegia canadensis*) has twice the nectar of the other columbines native to North America. The red flowers evolved to produce more nectar to satisfy the higher energy needs of hummingbirds compared to their other pollinators, such as bees and hawk moths.

Columbines came to America over the Bering land bridge, a broad land connection between Asia and Alaska during the Pleistocene era, some 10,000–40,000 years ago. From there, columbines spread throughout North America, evolving into new species. Just as with Darwin's finches, columbines developed different shapes and colors, in response to the new habitats and pollinators that they encountered.

There are twenty-one recognized species of native columbine in North America, but *A. canadensis* is the only columbine species native to eastern North America and is found throughout the range of the Ruby-throated Hummingbird, the only common hummingbird of the East.

Value to Birds and Other Wildlife

While hummingbirds and some bees visit columbine flowers for nectar and pollen, respectively, finches and buntings visit later in the year to eat the seeds. Some insects feed on the leaves, including the larvae of columbine duskywing butterfly, borer moths, the columbine sawfly, and several leaf miner flies. Mammals avoid the toxic foliage.

How to Grow

Wild columbine prefers soils that are moist in spring and dry later in summer when they go to seed and die back. It tends to be short-lived, but it self-sows, popping up in unexpected places with less competition and a patch of bare earth where their seeds can germinate. I discovered that ones I had planted outside my windows found a new home in an opposite corner of my yard, much to my surprise. Wild columbine, which is commonly found among rocky outcrops, is an excellent choice for rock gardens. Avoid overwatering, which can cause crown rot.

Propagation

Seeds; difficult to divide and transplant.

Companion Plants

Good choices to grow with wild columbine in a rock garden are Short's aster (*Symphyotrichum shortii*), harebell (*Campanula rotundifolia*), bulblet fern (*Cystopteris bulbifera*), zigzag goldenrod (*Solidago flexicaulis*), and smooth cliff brake fern (*Pellaea glabella*).

Wild columbine.
PHOTO BY ANNE MORETTI

A Closer Look

The spurs or tubes, formed from the petals of this delicate flower, evolved to the perfect length for the bill of the Ruby-throated Hummingbird. The spurs point upward and have a small bulge on the tip, which contains the nectar. Bees are sometimes attracted to columbine but can't reach the nectar in the spurs; instead, they may "rob" the nectar by poking a hole in the spur to sip the nectar. To detect these thieves, look for holes in the spurs of your columbines.

In late summer, check the leaves for light-colored trails, which are easy to spot. The narrow serpentine trails gradually widen, and you can also sometimes see dark spots within the trails. These trails are caused by two species of leaf miner flies (*Phytomyza aquilegiana*) and (*P. aqulegivora*), both specializing in columbines, as their Latin names suggest. The adult flies lay eggs on the underside of the leaves, and when the larvae hatch, they burrow between the layers of leaves and begin to eat their way through (thus the name *leaf miner*). As the larvae fatten up, the trails widen; the dark specks are their droppings. With luck, you might find a yellow-green larva at the end of a wide trail. When fully grown, the fly larva emerges and pupates in the soil.

TRUMPET CREEPER (*Campsis radicans*)

Trumpet Creeper Family, Bignoniaceae
Perennial Vine
Size: Height 6–40 feet, width 4–10 feet
Color and Season: Orange/red orange, yellow flowers in summer; yellow-green foliage in autumn
Light Requirement: Partial sun; sun
Soil: Moist to dry
Native Habitat: Woods, hedgerows
Native to: AL AR DE IA IL IN FL GA KS KY LA MD MO MS NC NJ OH OK PA RI SC TN TX VA WV
USDA Hardiness Zones: 4–10

This vine has a tropical air with flamboyant trumpet-shaped flowers up to three and a half inches long and a thicket of compound leaves on vigorous vines.

The trumpet creeper vine is a hummingbird magnet with the richest nectar of any North American plant.

PHOTO BY JACK BARTHOLMAI

The flowers produce nectar with 26 percent sugar content, exceptionally high for a temperate plant and considered the richest among North American flowers. The Latin name *Campsis*, from the Greek word *kampsis*, refers to the flower's curved stamens. *Radicans* derives from *radicant* (having root stems).

Value to Birds and Other Wildlife

Trumpet creeper's magnificent flowers are a magnet for hummingbirds and sphinx moths, both of which are effective pollinators. The flowers also attract Baltimore Orioles and Orchard Orioles, but these birds tend to tear up the flowers as they sip nectar. Trumpet creeper is host to seven caterpillar species, including those of the plebeian sphinx moth and the trumpet creeper moth. Many ants and flies, and some bees, are attracted to the extrafloral nectaries, which occur on the petioles, calyxes, corollas, and fruit of the vine. A leaf beetle feeds on the leaves. Mammals avoid the toxic foliage, which can give a rash to some people. The luxuriant foliage provides dense cover and nest sites for many songbirds.

How to Grow

This vine grows well in light shade but will blossom more profusely in sun and attract more hummingbirds. As mentioned in the introduction, trumpet vine can be quite aggressive, so careful thought should be given to whether and where to plant it. It can grow vigorously, as much as ten feet each year. To keep it within bounds, prune shoots each winter to about half a foot from the main trunk. Blooming won't be compromised because the plant flowers from new growth. The vine also sends out root sprouts that emerge some distance away from the trunk. The sprouts can usually be dealt with by mowing, but in unmowed gardens the sprouts will need to be dug out so they won't outcompete the other plants.

Despite these drawbacks, my husband and I grew trumpet creeper at our previous home with few problems. It was planted to camouflage the boring wall of our aluminum-sided garage, and we loved its showy flowers amid the vine's dark green leaves. The only problem we had was with its strong aerial rootlets, which left unsightly hard-to-remove marks on the siding in places where we had to cut back the vine's branches. We learned the hard way that it's best to use a sturdy trellis or other support several inches away from a wall on which the vine can grow. The rootlets can also damage wood, stone, and brick walls. We didn't have problems with root sprouts, as the vine was growing beside a concrete walkway that was edged with lawn. The vine can be an excellent choice to screen unwanted views; to climb and cover dead trees, brush piles, or fences; or to serve as a backdrop against walls.

In the right place, trumpet vine is an outstanding addition to a hummingbird garden. But some gardeners may prefer a less aggressive alternative, such as the native red honeysuckle vine (*Lonicera dioica*), another hummingbird favorite.

Propagation

Seeds or cuttings.

Companion Plants

Perennials, including some of the plants highlighted in this chapter (blazing stars, lilies, and bergamot), will look gorgeous against the lush green backdrop of the abundant foliage of this vine and will add irresistible allure to hummers.

A Closer Look

Trumpet creeper is spectacular at a distance, but it's extraordinary close-up too. If you examine the flowers, you'll often see ants crawling on them. A hand lens may help you discover why. Trumpet creeper has tiny pimple-like bumps on its flower parts—corolla, petiole, and calyx—and seedpods, which ooze tiny droplets of nectar. These are called extrafloral nectaries. The ants devour this nectar and attack and/or eat the thrips, aphids, and other insects that might threaten their sugary nectar supply. In this way, the ants serve as guards protecting the creeper's foliage and flowers. No other temperate zone plant has so many nectaries—a phenomenon usually found only in tropical or subtropical plants that have also evolved an ant-guard symbiosis.

In addition, the stigma of trumpet creeper flowers deserves close inspection. It is two-lobed and amazingly can open and close when pollinators visit. Research suggests that this is an adaption to enhance chances for pollination. After receiving some pollen, the stigma closes for a few hours, perhaps to resupply its nectar content, and later reopens to again invite pollinators. When enough pollen has been deposited, requiring 350 to 400 grains of pollen, the stigma closes for good so that the pollen is safely secured inside and won't be brushed off. After this, the fertilized flowers remain on the plant, all the better to maintain a showy floral display to continue to attract hummingbirds to the flowers not yet pollinated.

NEW JERSEY TEA (*Ceanothus americanus*)

Buckthorn Family, Rhamnaceae
Perennial Shrub
Size: Height 2–3 feet, width 2–3 feet
Color and Season: White flowers in summer, insignificant foliage color in autumn
Light Requirement: Sun to partial shade
Soil: Dry to medium; sandy, rocky
Native Habitat: Prairies, savannas, barrens
Native to: All states east of the Mississippi River plus AR IA KS MN MO LA NE OK TX; native but rare in DE ME
USDA Hardiness Zones: 3–9

New Jersey tea attracts a cornucopia of small insects that provide protein to supplement hummingbirds' sugary diet.

PHOTO BY SUSAN DAMON

This short, mound-forming shrub is covered with showy panicles of fluffy blossoms when flowering. Bloom lasts about three to four weeks. Unlike the flowers of the herbaceous nectar-rich hummingbird flowers, this shrub's flowers have a pleasant honey fragrance—obviously meant to attract insects, since hummingbirds use sight, not smell, to locate their preferred flowers. Inland New Jersey tea (*Ceanothus herbaceous*), a related shrub, is very similar in appearance and habitat preferences, but more rare over much of the range.

Value to Birds and Other Wildlife

New Jersey tea's flowers are a preeminent source of pollen and nectar, attracting droves of insects of all kinds. This cornucopia of insects, in turn, brings Ruby-throated Hummingbirds that snatch them up for high-protein meals to feed their nestlings and to supplement their sugary, carbohydrate-rich nectar diet. Among the amazing number of insects visiting New Jersey tea flowers in the Chicago region are more than thirty-five species of bees, forty species of beetles, twenty-five species of ants, eleven species of flies, eight species of bugs, at least two species of butterflies, and one skipper—all of which have been documented in *Flora of the Chicago Region*. Caterpillars of spring azure and summer azure butterflies and mottled duskywing skippers feed on the leaves. Throughout the shrub's range, it supports a total of forty-five species of butterflies and moths. Among the moth caterpillars are the broad-lined erastria, sulfur moth, and red-fronted emerald. Elk, deer, and rabbits browse on both the leaves and twigs. Turkeys and quail consume New Jersey tea seeds.

One fascinating observation of bees is reported in the *Flora of the Chicago Region*: "In a magnificent black oak savanna . . . the loud buzzing sound that emanates from thousands of *Bombylius fraudulentus* individuals as they hover above the flowers of New Jersey tea is amazing to witness if one can tolerate the more abundant mosquitos."

How to Grow

New Jersey tea grows best in full sun and moderately dry to dry soils. It is very drought resistant but will not survive in wetter soils. In urban/suburban areas, it is often devoured by rabbits and deer and will need to be protected, especially young transplants.

Propagation

Slow from seeds and cuttings. The seeds are explosive, so tie bags around clusters of seedpods to catch them.

Companion Plants

Pair with other low-growing shrubs that also do well in sun. I planted shrubby St. John's wort (*Hypericum prolificum*) and climbing prairie rose (*Rosa setigera*) near my New Jersey tea. Native shrub roses (*Rosa* species) and lead plant (*Amorpha canescens*) would also be good choices.

A Closer Look

Ceanothus species have a symbiotic relationship with nitrogen-fixing soil bacteria of the genus *Frankia*. The bacteria form nodules (actually galls) on the roots. The bacteria take nitrogen from the soil air and concentrate it into a form that can be used by plants. This is especially valuable for plants like New Jersey tea that grow in poor soils. Ultimately, the soil around New Jersey tea is enriched with some of this nitrogen, which can then be used by other plants as well.

I find it tricky to distinguish the leaves of invasive buckthorn (*Rhamnus cathartica*) shrubs from those of New Jersey tea when the plants are very small, perhaps not surprising since they are both in the buckthorn family. Both have leaves that can be alternate or opposite. When the two shrubs are full grown, it's easy to see their different forms. New Jersey tea grows into a short mound, while invasive buckthorn shrubs can be as tall as twenty-five feet. But don't wait until the invasive buckthorn grow taller and more difficult to eradicate. A few differences in their leaves can observed before the shrubs reach mature height, which can be helpful. Buckthorn's leaves are hairless and shiny and usually have a pointed tip. New Jersey tea leaves are soft and hairy, giving the shrubs a grayish cast, and they have a blunt tip.

ORANGE JEWELWEED (*Impatiens capensis*)

Touch-Me-Not Family, Balsaminaceae
Annual Wildflower
Size: 2–5 feet
Color and Season: Orange, red, yellow flowers in summer to early fall
Light Requirement: Full sun to shade
Soil: Moist to wet
Native Habitat: Woodlands, prairies
Native to: All lower 48 states except AZ CA CO MT NM NV UT WY; native but rare in FL
USDA Hardiness Zones: 3–9

Orange jewelweed is one of the last native nectar-rich hummingbird flowers to bloom in autumn. Ruby-throated Hummingbirds tend to follow the flower south each fall, devouring its rich nectar to fuel their mind-boggling migrations across the Gulf of Mexico. A close relative is yellow jewelweed (*Impatiens pallida*), which has a more limited range and tends to be visited more by bees than hummers.

The orange jewelweed's nectar-rich flowers help fuel the hummingbird's fall migration.

PHOTO BY JACK BARTHOLMAI

Jewelweed is well named, for the flower glistens like a pendant amber jewel on dewy sunny mornings. As well, the dew dripping from the pointed leaves sparkles in the light. Even submerged, the leaves have a lovely silvery sheen. It's also been discovered that the flowers fluoresce in ultraviolet light, revealing a chevron-shaped pattern. The chevron likely serves as a nectar guide for hummingbirds and most insects that can see into the ultraviolet range of light, unlike humans. Jewelweed's other name, spotted touch-me-not, is also fitting. *Spotted* comes from the many reddish-brown blotches on the flowers. When the seeds are ripe, a light touch or breeze causes the pods to fling their seeds up to four feet away—hence the name touch-me-not. This is fun for kids who love to snap the pods to see them explode and hear the seeds hitting nearby leaves.

The sap of the stems and leaves of jewelweed is said to relieve itching from poison ivy and stinging nettle.

Value to Birds and Other Wildlife

Jewelweed's nectar is a draw not only for hummingbirds but also for orioles and many bees, the latter often collecting the plant's pollen as well. The eastern tiger swallowtail butterfly and the hummingbird clearwing moth commonly visit to sip its nectar. The caterpillars of four other moths feed on the leaves, and several midges create galls on buds and leaves. Its large seeds are eaten by mice and game birds, including Ruffed Grouse, Ring-necked Pheasant, Greater Prairie Chicken, and Northern Bobwhite. The foliage provides cover for frogs and reptiles, and browse for deer.

How to Grow

Jewelweed seeds require two winter seasons to germinate. Perhaps the long wait for germination has given it the Latin name *impatiens* (impatience). But once established, jewelweed grows abundantly, self-seeding year after year. Not long ago, I collected jewelweed seeds from a large stand growing in abandon along a country road not far from home and strewed the seeds along the edges of a small pond in my yard. Jewelweed now rims our pond like a beautiful green necklace throughout the summer and sparkles with delicate orange blossoms.

Although jewelweed can be aggressive, it has shallow fibrous roots and is very easy to pull. Nonetheless, it may be difficult to manage in optimal conditions and should be closely monitored before it gets out of hand. On the positive side, dense stands of jewelweed have been found to compete successfully with the invasive garlic mustard (*Alliaria petiolata*), which threatens so many of our gardens and natural areas. The cultivated *Impatiens* found in garden stores is a non-native tropical species.

Propagation

Seeds; requires two seasons to germinate.

Companion Plants

Consider planting jewelweed with cardinal flower (*Lobelia cardinalis*), another hummingbird flower highlighted in this chapter that also does well in moist soils. The two would look stunning beneath pussy willow (*Salix discolor*), also included here, for an irresistible display to attract Ruby-throated Hummingbirds.

The orange jewelweed's nectar is contained in the spur curling beneath the floral pouch.
PHOTO BY ANNE MORETTI

A Closer Look

Take a peek in jewelweed flowers and you may see a beautiful metallic-green bee. The pure green sweat bee (*Augochlora pura*) is frequently found sleeping within its folds during the day. You may see another sign of bees by examining the spurs, which curl just below the floral pouch. Jewelweed's nectar is contained in these spurs, and bumblebees sometimes steal nectar by chewing holes in them. Later, smaller insects like syrphid flies visit the same holes to steal nectar.

Like many plants, jewelweed has a fascinating flowering strategy to increase successful pollination. There are two stages: the first is the male stage when the pollen is ripe, followed by a female stage when the stigma is sticky and ready to receive pollen. You can see which stage the flower is in by looking into the mouth of the flower. If the flower is in its male stage, you'll see white pollen. If it's in the female stage, you'll find a small green point—the stigma. Pollination happens when a hummingbird or large bee dives into a male stage flower for a sip of nectar and gets pollen on its body and later visits a flower in the female stage where the pollen is brushed onto its stigma. Within any population of jewelweed plants, there are flowers in both stages, ensuring pollination occurs readily.

BLAZING STAR OR GAY FEATHER (*Liatris* species)

Aster Family, Asteraceae

The various species of *Liatris* are surely among the most beautiful native flowers in North America. Commonly called blazing stars or gay feathers, they are favorites in any garden or prairie. They have many individual tubular flowers crowded together into a flower head. The flower heads have a fuzzy appearance due to the protruding styles (female flower parts) and stamens (male flower parts). There are two basic forms of *Liatris*. Some have flower heads closely packed together in slender plumes on the stems, resembling pink-purple feathers, giving them the name *gay feather*. Others have flower heads separated into button-like blooms spaced apart on the stems, giving them the apt name *blazing star* for their starburst look. The two common names are often used interchangeably.

Prairie gay feather and yellow coneflowers contrast beautifully in this Wisconsin prairie.

PHOTO BY MARIETTE NOWAK

Luckily for us, many species have evolved through adaptive radiation, diversifying into different forms to fill particular ecological niches with about forty species in all. Consequently, gardeners can easily find a species perfectly suited to the conditions in their gardens. For example, various species have adapted to different soils and habitats. Marsh gay feather (*L. spicata*) likes wetter soils and wet prairies, rough blazing star (*L. aspera*) does well in drier soils and dry prairies, while meadow blazing star (*L. ligulistylis*) is in the middle, preferring moist soils, neither too wet nor dry. Occasionally, some species grow relatively close to one another and frequently hybridize. Five of the most common and widely available *Liatris* species in the Midwest are featured here.

Value to Birds and Other Wildlife

Hummingbirds visit *Liatris* for nectar, while goldfinches, juncos, and many other birds eat the seeds. A host of insect pollinators, including butterflies, skippers, bees, beetles, and flies, visit for nectar. Bees also sometimes collect pollen for their larvae. Among the many butterflies, skippers, and moths that nectar on *Liatris* are the American and painted ladies, common buckeye, Dion skipper, fiery skipper, great spangled fritillary, giant swallowtail, Horace's duskywing, hummingbird clearwing, monarch, orange sulphur, Peck's skipper, sachem, silver-spotted skipper, snowberry clearwing, tawny-edged skipper, white-lined sphinx, wild indigo clearwing, and the non-native *cabbage white*. Two rare moth caterpillar species feed on *Liatris*. The caterpillars of the liatris flower moth feed on the florets and seed; caterpillars of the liatris borer moth bore into the stems.

How to Grow

All *Liatris* grow best in sun. They will get lanky if planted in shady locations or if overfertilized. Most can grow in a variety of soils and are drought tolerant, although different species will do best in their preferred conditions (see individual species treatment).

Propagation

Easy from seeds, but seedlings may be preferred because of the plant's slow growth. *Liatris* species usually grow from corm-like rootstocks, just below the soil surface, so be sure not to plant any deeper.

Companion Plants

The pink to purple flowers of *Liatris* are especially lovely in the garden when interspersed with white flowers like flowering spurge (*Euphorbia corollata*) or bright yellow coneflowers (*Ratibida pinnata*).

A Closer Look

Watch as *Liatris* begins to flower. Unlike most plants, it blossoms from the top down along the spike. Also, look carefully at a flower head and you'll see a tightly packed group of tubular flowers in each one. The flowers have a two-stage flowering sequence, a male stage and a female stage, similar to that of jewelweed and most asters. You can tell which stage the florets are in by the color of their anthers, the part of the male flower. If the anthers are deep purple, the fresh flowers are in the male stage; if they are tan, the flowers are in the later female stage. The deep purple color is a visual signal to pollinators, advertising that the florets are in the male stage and offer both nectar and pollen. Pollinators visit these male florets more frequently than female stage florets, which offer mainly nectar. Occasionally, *Liatris* has only white florets and pollinators rarely visit.

Although *Liatris* flowers may look similar, they can often be identified by the bracts, leaflike structures, just below the flower heads. The two plume-type *Liatris* species described here have either reflexed, pointed bracts like those of prairie gay feather or closely packed, blunt bracts like those of marsh gay feather. The three button-type *Liatris* featured also have different shaped bracts. The cylindrical blazing star has a short abrupt tip on its bract. The other two have rounded bracts. The length of the flower stalks can also differ. For example, meadow blazing star has long stalks on its flower heads, whereas rough blazing star has either no stalks or very short ones.

Rough Blazing Star (*Liatris aspera*)

Perennial Wildflower
Size: 2–5 feet
Color and Season: Pink to purple flowers in July to October
Light Requirement: Sun
Soil: Dry to moderate moisture; loam, sand
Native Habitat: Prairies
Native to: AL AR FL GA IA IL IN KS KY LA MI MN MO MS ND NE OH OK SC SD TN TX VA WI WV; native but rare in NC
USDA Hardiness Zones: 3–8

Rough blazing star is one of the button-type species of *Liatris*. The name *aspera* (rough) reflects the rough-surfaced leaves of the species. The plant is

The rough blazing star, one of the last *Liatris* species to blossom in late summer, attracts many butterflies, like this great spangled fritillary.
PHOTO BY MARIETTE NOWAK

tough and drought resistant. In mid to late August, glorious displays of these "stars" can be found in dry prairies on the ancient moraines left by the glaciers where I live. This species also does well in the dry gravelly soils in my small yard.

Cylindrical Blazing Star (*Liatris cylindracea*)

Perennial Wildflower
Size: 8–24 inches
Color and Season: Pink to purple flowers in summer to early fall
Light Requirement: Sun
Soil: Dry; poor soil, tolerates loamy well-drained soil
Native Habitat: Prairies
Native to: AR IA IL IN MI MN MO WI; native but rare in AL GA KY OH TN NY
USDA Hardiness Zones: 4–7

This lovely button-type *Liatris* is the species most adapted to poor soil and dry conditions. It is shorter than most other *Liatris* and begins to blossom earlier, starting in July. It often grows on rocky hillsides but will tolerate loamy well-drained soil. I planted them in my yard on a dry slope.

Meadow Blazing Star (*Liatris ligulistylis*)

Perennial Wildflower
Size: 3–5 feet
Color and Season: Pink to purple flowers in summer to early fall
Light Requirement: Sun
Soil: Medium to moist; loamy soils
Native Habitat: Prairies
Native to: CO IA IN MN MO MT ND NM SD WI WY; native but rare in NE
USDA Hardiness Zones: 3–6

A button-type *Liatris*, this species is the ultimate magnet for monarch butterflies, attracting more than any other *Liatris* and indeed, more than any other flower I know. Like all *Liatris* plants, it is also visited by Ruby-throated Hummingbirds and numerous other butterflies and pollinators. Its seeds are devoured by goldfinches in autumn. The flowers tend to bloom all at the same time and are showy for weeks in late summer. Although the plant can handle rather dry soil,

This giant swallowtail is nectaring on a meadow blazing star.
PHOTO BY MARIETTE NOWAK

it is best planted in moist soils. I have only a few in a fairly moist spot in my prairie. But soils that stay wet in winter should be avoided, as they can harm this species.

Prairie Gay Feather (*Liatris pycnostachya*)

Perennial Wildflower
Size: 3–5 feet
Color and Season: Pink to purple flowers in July–September
Light Requirement: Sun
Soil: Medium to moist; clay, loam, sand
Native Habitat: Prairies
Native to: AR IA IL KS KY LA MN MO MS ND SD OK TX WI; native but rare in NE IN
USDA Hardiness Zones: 3–9

This species has beautiful plumes of purple flowers. The species name, *pycnostachya*, means dense or thick and refers to its crowded flower heads and leafy stems. Prairie gay feather grows well in moist garden soils and is one of the most popular and widely planted *Liatris* species. Those in my garden are doing well and bloom profusely, especially after a prescribed burn. Prescribed burns mimic the natural fires caused by lightning or those set by American Indians in the past and are essential for managing prairies and savannas.

Marsh Gay Feather (*Liatris spicata*)

Perennial Wildflower
Size: 3–6 feet
Color and Season: Pink to purple flowers in summer to early fall
Light Requirement: Sun
Soil: Medium to moist; clay, loam, sand
Native Habitat: Prairies
Native to: AL CT FL GA IA IN KY LA MI MS NJ OH PA TN VA WV; rare in WI MD; extirpated in DE MO
USDA Hardiness Zones: 4–10

Marsh gay feather has tall, showy plumes on leafy stems. Unlike most species of *Liatris*, it prefers damper soils but will do well in average garden soils. Watch for a camouflaged caterpillar (*Synchlora aerata*) nestled in the flowers. This caterpillar covers its body with bits of the flowers to hide as it munches on the petals.

MICHIGAN LILY (*Lilium michiganense*)

Lily Family, Liliaceae
Perennial Wildflower
Size: 3–7 feet
Color and Season: Orange flowers in summer
Light Requirement: Sun to partial shade
Soil: Moist to wet; loam, sandy loam
Native Habitat: Prairies, woodlands
Native to: AR IA IL IN KY MI MN MO OH WI; native but rare in GA KS NE NY OK SD TN
USDA Hardiness Zones: 3–7

A spectacular wildflower, Michigan lily has all the characteristics of an ideal hummingbird flower magnified in size. The pendant petals of this large two- to three-inch flower are orangish red dotted with purple spots and curve back

The gorgeous Michigan lily has all the attributes of a perfect hummingbird flower, from its color to its structure.
PHOTO BY ANNE MORETTI

nearly to the base of the flower, hiding deep nectar tubes. Six large stamens and a single pistil, almost the length of the corolla itself, project below, offering an irresistible invitation to hummers and other pollinators seeking the rich nectar within.

There are several closely related native species. The Canada lily (*Lilium canadense*) is so close a relative and look-alike that it has been proposed that the two are the same species. Turk's-cap lily (*L. superbum*), formerly considered the same species as Michigan lily, is now recognized as a separate species. Its native range is more southeasterly, and it can be identified by the green star pattern usually present at the base of its flowers. Prairie lily (*L. philadelphicum*) has upright flowers and is less widely available commercially.

Value to Birds and Other Wildlife

Hummingbirds and orioles are both attracted to the showy flowers. Larger moths, such as hummingbird and sphinx moths, come to nectar, as do large butterflies, including the fritillaries, monarchs, and swallowtails. The metallic-green bee *Augochlora pura* also visits the flowers for nectar. Some beetles and moth larvae feed on the stems and bulbs of these lilies.

Unfortunately, deer love the lilies and devoured the first lilies I planted. I protected my second planting of these beauties with a green chicken-wire fence, not too noticeable amid the green leafy vegetation around the plants. Other mammals sometimes browse on the plants, and voles may eat their bulbs.

How to Grow

A prize for any garden, Michigan lily grows well in average moist garden soil. It is an excellent choice for rain gardens and low areas in perennial gardens. But it can be a bit fussy about growing conditions. Mulch to keep the soil cool and moist. Avoid too much fertilizer and watering, which will cause the plants to flop. Since the plant is a favorite of deer, you may need to cage it; be sure to make the cage wide enough to allow the plant to spread. On our former small property, we had no deer and grew these lilies for years. But deer are abundant where we now live, and we quickly lost our lilies to these browsers. If deer are abundant where you live and your growing conditions for Michigan lily are not ideal, you may want to consider growing its sister plant, great blue lobelia

(*Lobelia siphilitica*), which is easier to grow and not as vulnerable to deer. Other excellent alternatives for moderately most soils are wild blue phlox (*Phlox divaricata*) and foxglove beardtongue (*Penstemon digitalis*). All three of these species have nectar-filled flowers sure to attract hummingbirds.

Propagation

Challenging from seeds. Easier and faster to grow from transplants or bulbs.

Companion Plants

Plant with other moisture-loving native species, such as sneezeweed (*Helenium autumnale*), turtlehead (*Chelone glabra*), royal fern (*Osmunda regalis*), and joe-pye weed (*Eutrochium maculatum*).

A Closer Look

The sepals and petals of Michigan lily look the same and are commonly referred to as tepals. Take a look at the number of flower parts. Like most lily family species, they are all in threes or multiples of three. Note the six tepals in two whorls of three and the six stamens. There is a single pistil, but its ovary is divided into three compartments and after pollination matures into a three-celled seedpod. You can also see that the leaves are parallel veined, another characteristic of plants in the lily family.

CARDINAL FLOWER (*Lobelia cardinalis*)

Bellflower Family, Campanulaceae
Short-Lived Perennial Wildflower
Size: 2–5 feet
Color and Season: Scarlet flowers in late summer
Light Requirement: Sun, light shade
Soil: Moist to wet; loam, sand
Native Habitat: Woodlands, stream banks
Native to: All lower 48 states except ID ND MO OR SD WA WY; native but rare in CO NE
USDA Hardiness Zones: 3–9

The bright red cardinal flower is a hummingbird favorite.
PHOTO BY MARIETTE NOWAK

These breathtaking, intensely red flowers are favorites of hummingbirds. I was thrilled when a few showed up along the edges of my little kettle pond a few years ago after disappearing previously. The common name refers to the red robes of Roman Catholic cardinals. The plants blossom for weeks as their buds gradually open in sequence up the stem. In late September the final top blossoms resemble glowing crimson flames atop candle-like stems.

Value to Birds and Other Wildlife

The flowers are very attractive to hummingbirds and swallowtail butterflies, but not Northern Cardinal songbirds, despite their shared name. Swallowtails can detect the color red, unlike bees. Nonetheless, bumblebees sometimes steal nectar through slits in the nectar tubes, and halictid bees gather the flower's pollen. The larva of the fly (*Metopomyza scutellata*) "mines" the insides of leaves. Pink-washed looper moth caterpillars feed on the leaves. Mammals avoid the foliage due to the toxic latex sap.

How to Grow

Cardinal flowers grow well in moist perennial gardens and rain gardens. They also can be used to stabilize stream banks. In any setting, their roots need to be kept moist at all times. Cardinal flowers are short-lived perennials that don't persist more than a few years. But they produce hundreds of tiny seeds, which can repopulate an area as long as the seeds have contact with the soil and little competition. Rosettes of leaves are also formed near the base of the parent plant and need to be kept free of competition in order to grow roots before winter. Mulching will keep the rosettes moist and reduce frost upheavals. For a more long-lived perennial, you may want to consider growing its sister plant in the same genus, great blue lobelia (*Lobelia siphilitica*), which also grows in moist to wet soils. Other excellent alternatives for moderately moist soils are wild blue phlox (*Phlox divaricata*) and foxglove beardtongue (*Penstemon digitalis*). All three of these species have nectar-filled flowers that are very attractive to hummingbirds.

Propagation

Seeds or transplants.

This female Ruby-throated Hummingbird is probing the deep tube of a cardinal flower for nectar.

PHOTO BY DAWN-MARIE STACCIA

Companion Plants

Group with other wetland natives, like bright yellow sneezeweed (*Helenium autumnale*), white turtlehead (*Chelone glabra*), and blue flag iris (*Iris virginica*). The flowers also look spectacular against a dark backdrop of ferns or shrubs.

A Closer Look

This species has a two-phase flowering in which the male and female parts of the flowers become receptive to pollinators at different stages. Nectar is produced in both phases so pollinators are sure to visit both. To check the flowering phases, look for the narrow tube at the top of the flowers. It consists of five anthers joined to one another. During the male phase, two of the anthers have fuzzy white hairs, which help cause the anthers to dust visiting pollinators with pollen. In the female phase, a small female stigma in the shape of an open clamshell, projects from the anther tube. This stigma scrapes against the pollen carried by a pollinator that previously visited a male flower. An ingenious method to ensure pollination!

WILD BERGAMOT (*Monarda fistulosa*)

Mint Family, Lamiaceae
Perennial Wildflower
Size: 2–5 feet
Color and Season: Pale purple to pink flowers in summer
Light Requirement: Sun, part shade
Soil: Moist to dry; clay, loam, sand
Native Habitat: Prairies, savannas
Native to: All lower 48 states except CA FL NV; native but rare in DE; extirpated in RI
USDA Hardiness Zones: 3–9

Wild bergamot's showy clusters of flowers resemble pink pompoms in the prairie.
PHOTO BY WENDY GOCHENAUR

This popular garden plant, native throughout most of the country, has loads of lovely raggedy-looking pompoms of blossoms. The name *bergamot* refers to a citrus tree (*Citrus bergamia*) grown in Italy that has a similar fragrance. Earl Grey tea is flavored with the oil from this citrus tree. I'm fortunate to have wild bergamot flowers in two different hues in my small prairie—pale purple as well as rich pink. Red beebalm (*Monarda didyma*) is a closely related species, but native only to easternmost states, although widely planted throughout the Midwest.

Value to Birds and Other Wildlife

Wild bergamot is an outstanding source of nectar for hummingbirds and many species of butterflies. The nectar is available continually through the day, as new flowers open and older ones are depleted. Hummingbird moths, long-tongued bees, and a variety of other insects also nectar on the flowers. The small black bee *Dufourea monardae* specializes in wild bergamot, visiting for both nectar and pollen. Caterpillars of the hermit sphinx and gray marvel moths feed on the leaves. Deer and other mammals usually avoid this plant, perhaps because of the citrus/oregano flavor of the leaves. Sparrows, goldfinches, and redpolls eat the seeds.

How to Grow

Wild bergamot is common in prairies and savannas on all but the wettest of soils. It's a showy plant for prairie and perennial gardens and can be used for slope and buffer stabilization. Fertilizing and overwatering can cause it to flop. Wild bergamot is susceptible to powdery mildew, but this is rarely a serious problem and can be reduced by refraining from overhead watering. The plants can be divided every few years to sustain their vigor.

Propagation

Seeds or root division.

Companion Plants

The pink/purple flowers contrast beautifully with the many yellow flowers blossoming at the same time—black-eyed Susan (*Rudbeckia hirta*), yellow coneflower (*Ratibida pinnata*), prairie coreopsis (*Coreopsis palmata*), and others.

A Closer Look

First, feel the stems of wild bergamot. They are square, a characteristic of plants in the mint family. Next, examine the rounded heads of wild bergamot, which contain many individual tubular flowers. The species name of wild bergamot, *fistulosa*, refers to the tubelike flowers. These flowers bloom in sequence. Those in the center bloom first, and gradually other flowers begin to bloom toward the outer flower head. Eventually you will see a wreath of fresh flowers around the periphery of the flower head.

Each individual flower has a tubular upper lip with two stamens and three thin lower lips. The stamens produce their pollen first; later the female two-part stigmas open up to receive pollen. You may find holes at the base of the floral tubes where the nectar is located. These are made by our largest carpenter bee, *Xylocopa virginica*, to steal nectar, bypassing the floral tube opening. By doing so, it does not contact the pollen, and pollination cannot occur when it visits other flowers. Several other bees use the holes created by *Xylocopa* to

A close look at the individual flowers in a wild bergamot cluster reveals their lovely, intricate structure.

PHOTO BY MARIETTE NOWAK

pirate nectar for themselves. You may also see translucent yellow aphids (*Aphis monardae*) on the flowers tended by ants. And look in the foliage for a beelike robber fly (*Laphria thoracica*) that lies in wait for its insect prey to fly by. What a miniature menagerie of wildlife can be discovered on a single species like wild bergamot!

PUSSY WILLOW (*Salix discolor*)

Perennial Shrub, Willow Family, Salicaceae

Size: Height 6–15 feet, width 4–8 feet
Color and Season: Yellowish-gray or green catkins in late winter to early spring; yellow-green foliage in autumn
Light Requirement: Sun, part shade
Soil: Moist to wet; clay, loam, sand
Native habitat: Wetlands, marshes
Native to: CT IA IL IN MA ME MI MN MT ND NH NJ NY OH PA RI SD VT WI WY; native but rare in CO DE ID MD MO OK VA WV; extirpated in TN
USDA Hardiness Zones: 4–8

Along with the Red-winged Blackbird, the beloved pussy willow is a harbinger of spring. The male catkins are covered with gray silky fuzz (pussy fur), which gives the plant its common name and protects the catkins from cold. The species name *discolor*, meaning two or different colors, refers to the leaves, which are green above and powdery white below. The male flowers soon emerge from the silky fuzz, revealing masses of stamens tipped with golden pollen. The female flowers, much more subdued, have slim green pistils. Willows are dioecious, with all male or all female flowers on separate shrubs. Only the male plant produces pollen, but both sexes produce nectar. Willow bark contain salicin, a key ingredient in aspirin, which was developed from salicin in the 1800s.

Value to Birds and Other Wildlife

Pussy willow flowers attract swarms of small insects for hummingbirds that offer high-protein meals for adults and nestlings alike. The cottony hairs on the pussy willow seeds are used by hummers for nesting material.

Pussy willow's flowers attract small insects that nourish hummingbirds.
PHOTO BY MARIETTE NOWAK

An amazing diversity of bees visit pussy willow flowers for pollen. Forty-five species were observed in the Chicago region alone, including a few of which are specialists on *Salix* pollen. Butterflies visiting for nectar include the eastern comma, mourning cloak, question mark, red admiral, and spring azure. The small, colorful ailanthus webworm moth also feeds on the nectar.

The willow genus, as a whole, attracts more than 450 species of butterfly and moth caterpillars that feed various parts of plants. Many other insects feed on the leaves and stems, including leaf beetles, aphids, thrips, plant bugs, stink bugs, and the larvae of some sawflies—all providing a bounty for birds. In addition, the early season seeds of willows offer another food source for songbirds and waterfowl. Ruffed Grouse, White-crowned Sparrows, squirrels, turtles, rabbits, muskrat, beavers, deer, and elk all feed on their favored willow offerings from buds and branches to foliage and seeds. The American Goldfinch, Gray Catbird, Northern Harrier, Warbling Vireo, Willow Flycatcher, Wilson's Warbler, and Yellow Warbler often construct nests hidden within willow foliage.

How to Grow

Pussy willows, like other willow shrubs, are easy to grow and require little care. They should be cut back heavily every few years to encourage new growth. This can also help limit the spread of the shrubs when desired, since they sucker vigorously in moist soil. If planting young nursery plants, install three to five plants to ensure the presence of both female and male plants, which are separate in willows. Both sexes are needed to cross-pollinate and produce seeds.

Propagation

Willows contain a natural rooting hormone and are very easy to root from stem or root cuttings, either in moist ground or in water. Moderately easy from seeds.

Companion Plants

Shrubs that thrive in moist soil can be grown with pussy willow to create a shrubbery or windbreak that will be very attractive to many species of birds. Ninebark (*Physocarpus opulifolius*), pagoda dogwood (*Cornus alternifolia*), red osier dogwood (*Cornus stolonifera*), and serviceberry (*Amelanchier canadensis*) are excellent choices.

The caterpillars of the handsome mourning cloak butterfly feed on willow leaves.

PHOTO BY MIKE REESE, wisconsinbutterflies.org.

A Closer Look

The mourning cloak is a beautiful butterfly, with rich brown wings accented with bright blue spots and edged in yellow. Unlike many other butterflies, it hibernates as an adult under leaves or tucked in tree bark during the winter, enabling it to be one of the first to fly in spring. Watch for it and you may see a female laying her eggs on a willow, one of the butterfly's favorite plants. The eggs are laid in clusters around willow twigs. When the eggs hatch, look for clusters of the black-and-red caterpillars that eat the leaves voraciously, sometimes stripping the willow of foliage—but rarely causing any long-term damage to the willows.

Also, check pussy willow for what appear to be bird droppings but likely are the caterpillars of viceroy butterflies, another species that favors shrubby species of willow. The "poop" look helps protect them from predators. As adults, viceroys resemble monarch butterflies but have a black line across the hind wings. In the past, it was thought that birds avoided eating viceroys because they looked like monarchs, which contain toxins, but recently it's been discovered that viceroys, at least in some areas, produce their own toxic chemical. You're also sure to find many other insects visiting pussy willow, given the astonishing number attracted to this attractive willow.

ROYAL CATCHFLY AND FIRE PINK (*Silene* species)

Pink Family, Caryophyllacea

The *Silene* genus has many introduced non-native species, some of which are invasive. Here in North America, they have few of the insects and diseases that keep them in check in their home countries. But there are also a number of native species, including two outstanding hummingbird flowers with brilliant red colors and rich nectar: fire pink (*S. virginica*) and royal catchfly (*S. regia*). Unfortunately, both are more rare than the introduced species and have sometimes even been extirpated locally. I myself haven't seen either of them in the wild, although I've seen them in restored native gardens. Both are available at native plant nurseries, so consider planting them in your yard, if you have suitable location, to help preserve these beautiful flowers for future generations of hummingbirds as well as for us humans. Gardeners can be guardians of biodiversity right on their own properties by planting and protecting native species. As Aldo Leopold has said: "A plant must be saved in *many places* if it is to be saved at all."

Value to Birds and Other Wildlife

Both flowers are favorites of Ruby-throated Hummingbirds, who avidly sip their abundant nectar. Large butterflies like black swallowtails are also known to visit. Seeds attract juncos, Pine Siskins, sparrows, Water Pipits, and Horned Larks.

The rare royal catchfly has a green nectar tube below its crimson-colored petals.
PHOTO BY SUSAN DAMON

How to Grow

The two species differ in their basic preferences. Fire pink prefers the dappled shade of drier savannas, whereas royal catchfly does best in moist, sunny prairies. See details for each in the species descriptions. Both are relatively short-lived perennials but self-sow readily.

Propagation

Seeds; self-sows.

Companion Plants

These two species, being relatively short, do well with short grass prairie plants, like little bluestem (*Schizachyrium scoparium*), prairie dropseed (*Sporobolus heterolepis*), and black-eyed Susan (*Rudbeckia hirta*).

A Closer Look

A common name for members of the *Silene* genus is catchfly, which refers to the sticky hairs on the plants. The hairs act like flypaper to deter ants and other small insects from feeding on the plant. Unlike the pitcher plant (*Sarracenia purpurea*), which grows in nutrient-poor bogs, *Silene* species don't feed on the insects. The stickiness is simply a defensive strategy. However, hummingbirds eat these flies, another food source for hummers besides the rich nectar offered by the flowers.

Another strategy of *Silene* species is their staggered development of the stigmas and anthers, which flower at different times, preventing any given flower from self-pollination.

Royal Catchfly (*Silene regia*)

Perennial Wildflower
Size: 2–3 feet
Color and Season: Crimson flowers in summer
Light Requirement: Sun to part shade
Soil: Moist; loam

Native Habitat: Prairies
Native to: MO; native but rare in AL AR FL GA IN OH OK WI; extirpated in KS TN
USDA Hardiness Zones: 5–9

Extremely rare due to the loss of prairie habitat in the Midwest, this species has been extirpated in several entire states. It is one of the few prairie plants that are red and, unsurprisingly, one of the few prairie plants pollinated by hummingbirds. It's relatively easy to grow and should be planted more frequently in prairie gardens for its value for hummingbirds, for its outstanding beauty, and for conservation of this endangered species.

Fire Pink (*Silene virginica*)

Perennial Wildflower
Size: 6–14 inches
Color and Season: Crimson flowers in spring to early summer
Light Requirement: Partial shade, shade
Soil: Dry. moderately dry; loam, sand
Native Habitat: Savannas, open woods
Native to: AR GA IL IN KY LA NC MD MO MS OH OK PA SC TN VA WV; native but rare in AL DE FL MI TX WI; extirpated in KS
USDA Hardiness Zones: 4–9

The name *fire* is based on the fiery red color of its flowers. The additional name *pink* is a bit confusing. It refers not to the color but to the pink family to which the plant belongs. Fire pink has a wide native range but is relatively uncommon and even rare in some locales. It is rather fussy about its light preferences. Fire pink does not do well in bright sun or in heavier shade, either of which can lead to decline. But when planted in the dappled shade typical of the savannas to which it is native, fire pink is a stunning addition to perennial gardens.

An American Goldfinch probes for insects in a purple coneflower.

PHOTO BY DAWN-MARIE STACCIA

5 Top 10 Plants for Goldfinches and Other Prairie Birds

So we passed on in wonder until a prairie burst in to view and the wagon could hardly contain us for among the waving grass were millions of flowers, flowers such as we had never seen before. It was indeed a land of wonder and beauty and our eyes feasted on the *sights*.

—Isaphena Burdick, 1841, Walworth County, Wisconsin

What an amazing sight the original prairies once were! Many pioneers and early explorers expressed awe at their beauty and vast reaches. Tragically, more than 99 percent of the original 142 million acres of prairie in North America have been lost to plow, cow, and urban sprawl. Large-scale prairie restorations are helping to recover some of this lost landscape. But we can also reproduce small pocket prairies on our own properties and bring back a portion of prairie life—birds, bees, butterflies, and more—right in our own yards, whether small or large.

Prairies are sunny, treeless, natural communities with a rich diversity of gorgeous wildflowers, grasses, and wildlife. This chapter focuses on the top plants for birds in two of the main families of prairie species, the composite and the grass families. These plants offer nectar for hummingbirds, support an abundance of protein-packed insects for birds, and produce nourishing seeds that are rich in fats, proteins, and vitamins for migrant and resident birds alike. Together they provide the perfect diet for attracting and sustaining prairie birds.

The composite family is often also referred to as the daisy or aster family. "Flowers" in the composite family are actually clusters or composites of many small flowers contained in one flower head. They usually have two types of flowers within the flower head. The petals are called the ray flowers because

Small pocket prairies support songbirds, butterflies, and bees.
PHOTO BY MARIETTE NOWAK

they radiate out from the center of the flower head. The center of the flower head is packed with tiny flowers called disk flowers. Composite plants attract an astonishing variety of insects. Goldenrods, asters, and sunflowers, all highlighted in this chapter, are the top three groups of herbaceous plants for their role in supporting insect biodiversity, the basis of a thriving food web. Because of this, they are considered keystone species, essential for healthy ecosystems. Other composites covered here are coneflowers, native thistles, and silphiums.

Grass family plants have flowers, too, but they often go unnoticed. Most lack showy petals and sepals because they are wind pollinated and have no need to attract insect pollinators. However, their tiny intricate flowers often have lovely hues of gold or red and are well worth a close look with a hand lens. More importantly, their seeds are nutritious food for birds and other wildlife. Three species in this chapter are grasses. Prairie dropseed is a relatively short grass, ideal for pocket prairies, but widely planted as well in larger restorations. (Another short grass species, little bluestem grass [*Schizachyrium scoparium*], described in chapter 6, is an excellent choice for short grass prairies as well as savanna landscapes.) Switchgrass and Indian grass are taller grasses common in tall grass prairies but often used in smaller sites when feasible.

Common Birds in Home Prairies

These birds commonly spend at least some time during their life cycle—from foraging and roosting to courtship and nesting—in small prairie gardens and occasionally in larger prairies as well.

Red-winged Blackbird
Northern Cardinal
Gray Catbird
Black-capped Chickadee
Brown-headed Cowbird
American Crow
Mourning Dove
American Goldfinch
Ruby-throated Hummingbird
Blue Jay
Baltimore Oriole
American Robin
Chipping Sparrow
Song Sparrow
Tree Swallow
Eastern Towhee
Cedar Waxwing
Downy Woodpecker
House Wren

A Black-capped Chickadee forages among sunflowers (*Helianthus* species) searching for some of the 65 species of caterpillars that feed on the plants.
PHOTO BY JACK BARTHOLMAI

Major Prairie Birds in the Midwest

These birds require treeless prairies for most or all of their breeding cycle. Some are rare and endangered as a result of the loss of the vast original prairies.

Bobolink
Dickcissel
Northern Harrier
Horned Lark
Eastern Meadowlark
Western Meadowlark
Short-eared Owl
Greater Prairie Chicken
Yellow Rail
Upland Sandpiper
Grasshopper Sparrow
Henslow's Sparrow
Le Conte's Sparrow
Savanna Sparrow
Sharp-tailed Sparrow
Vesper Sparrow
Sedge Wren

Eastern Meadowlark.
PHOTO BY JEREMY MEYER, jeremymeyer.photography

NATIVE THISTLE (*Cirsium* species)

Aster or Composite Family, Asteraceae or Compositae

> With sublime blue-green foliage, interesting stem and leaf architecture, and pink blossoms, our native thistles are every bit as resplendent as countless other native plants.
>
> —James Eckberg et al., *Native Thistles*

It's time to bring back the native thistles. This may surprise you because most of us know only the invasive thistles that can take over en masse in many areas like an invading army. Native thistles have been getting a bad rap because of these thugs. But there are some beautiful native thistles that are very well behaved and invaluable to birds, butterflies, and other wildlife. Unfortunately, many native thistle species are in decline, and to stem their loss, they should be planted more often in our gardens and restorations.

Distinguishing native thistles from non-native thistles is an important first step in conserving our native species. Unfortunately, there are no absolute features of the plants that can separate all non-native thistles from native thistles. All thistles are spiny, but native thistles do not have spiny winged stems as do the common invasives—bull (*Cirsium vulgare*), spiny plumeless (*Carduus acanthoides*), and nodding thistles (*Carduus nutans*). However, one of the worst invasive thistles, Canada thistle (*Cirsium arvense*), does not have spiny wings. Consult a good identification guide, such as the online Field Guide to Terrestrial Invasive Plants in Wisconsin, to learn to identify these bullies.

The lavender blossoms of the native prairie thistle attract American Goldfinches and a variety of butterflies.
PHOTO BY MARIETTE NOWAK

Thistles are in the aster family, and like all species in the aster family, they are composites with dozens of individual flowers within a single flower head. Many composites, like sunflowers, have flower heads with several petal flowers surrounding a central disk with numerous disk flowers. Thistle flower heads, however, are composed entirely of disk flowers, each of which has both male and female parts.

Two native thistles in the *Cirsium* genus, available at native plant nurseries, will be described here: prairie or field thistle (*C. discolor*) and swamp thistle (*C. muticum*). Both are biennial, growing a rosette of leaves the first year and then sending up a flowering stalk the following year from the center of the rosette, forming seeds and then dying at the end of the growing season.

Value to Birds and Other Wildlife

Ruby-throated Hummingbirds often nectar on native thistles, some of which are known to have a higher nectar content than the flowers of many other species. A wide diversity of birds, including the American Goldfinch, Indigo Bunting, Pine Siskin, Clay-colored Sparrow, and other sparrows feed on the seeds of native thistles, many of which are known to be highly nutritious. American Goldfinches, in particular, appear to have a special relationship with thistles, delaying their nesting until thistle seed is available (see more later in this section).

Many butterflies visit the flowers of native thistles more frequently than those of most other plants, sometimes even when there are fewer thistles than

The American Goldfinch feeds voraciously on thistle seeds.
PHOTO BY JACK BARTHOLMAI

the other species. The diversity of butterflies, skippers, and moths attracted to the flowers include the black dash, black swallowtail, Delaware skipper, eastern tiger swallowtail, giant swallowtail, fiery skipper, great spangled fritillary, hummingbird clearwing, monarch, painted lady, pearl crescent, Peck's skipper, red admiral, the rare regal fritillary, sachem, silver-spotted skipper, snowberry clearwing, tawny-edged skipper, and white-lined sphinx. Caterpillars of the painted lady butterfly feed on the leaves of prairie thistles, while swamp metalmark caterpillars prefer swamp thistles.

A host of other native insects, including endangered species and thistle specialists, feed on the flowers, leaves, seeds, and/or stems of native thistles. Among them are ants, aphids, bees, lady beetles, and many other beetles, seed bugs, milkweed bugs, fruit flies, grasshoppers, treehoppers, wasps, and weevils. In the Midwest, at least eleven species of bees, including the endangered rusty patched bumblebee (*Bombus affinis*), feed on the rich nectar. Referring to the rusty patched bumblebee, Susan Carpenter, native plant garden curator at the University of Wisconsin Arboretum, reports, "We often find two or three bumblebees on the flower head," a wonderful observation given the bee's rare status. Cavity-nesting bees favor thistle stalks for nesting sites.

How to Grow

Thistles are best grown in the habitats to which they're adapted. Prairie thistle grows best in prairies or meadows with full sun in medium-moist soils. Swamp thistle, as its name suggests, prefers wetlands in full sun. These thistles can be beautiful additions to any landscape and you can choose the one best suited to the conditions on your property. By growing them, you'll be helping to conserve species that are invaluable to a multitude of birds, butterflies, and other wildlife species.

Propagation

Seeds; self-seeds readily.

Companion Plants

Choose native plants adapted to the same habitat as the thistle species you choose to grow. Prairie thistle grows well with prairie species such as flowering spurge (*Euphorbia corollata*), yellow coneflower (*Ratibida pinnata*), and big bluestem grass (*Andropogon gerardii*). Swamp thistle is beautiful with wetland species like turtlehead (*Chelone glabra*), the native narrow-leaved loosestrife (*Lysimachia quadrifolia*), and marsh blazing star (*Liatris spicata*).

A Closer Look

American Goldfinches are well worth observing throughout the year. Males perch high and sing their beautiful warbling melodies in spring, but nesting begins first in late June or early July, later than most other songbirds. It is timed, seemingly, to the availability of one of their favorite foods—thistle seed. Goldfinches feed voraciously on the seeds, possibly because they can feed more efficiently on thistles than on many other flowering species. They also use thistle down, the white silky fibers in the seed heads, to line their nests. In addition, goldfinches are one of relatively few bird species that feed seeds, rather than insects, to their nestlings. As a result, nutritious seeds like those of thistles are critical for their successful breeding.

The American Goldfinch lines its nest with thistle seed fibers.
PHOTO BY JACK BARTHOLMAI

Caterpillars of painted lady butterflies commonly feed on thistle foliage.
PHOTO BY MIKE REESE, wisconsinbutterflies.org

Thistles are also a favorite of painted lady butterflies, which prefer to lay their eggs on the plants. After hatching, the caterpillars make silken tents on the top of leaves where they feed, one caterpillar per tent. Look for the caterpillar tents when the thistle flowers are about to open. The frass (bug poop) collects on the bottom of the tents and often makes them easier to find. You may be able to see the caterpillars right through the loosely woven tents, but sometimes they leave their tents to feed on nearby leaves. The one-and-a-half-inch caterpillars come in a variety of colors but always have a black line along their backs and a yellow, often broken, stripe along each side. When they have eaten their fill, the caterpillars hang upside down from a leaf on a silken thread and form a chrysalis. The beautiful orange-and-black painted lady butterflies that emerge sip nectar from thistles and other flowers. Those in the northern United States migrate to our southern states and Mexico to winter. Their flight back in spring can be impressive when conditions are right, with great swarms of the butterflies returning to northern climes.

Prairie or Field Thistle (*Cirsium discolor*)

Biennial or Short-Lived Perennial Wildflower
Size: 3–10 feet
Color and Season: Pink to lavender (rarely white) flowers in mid to late summer
Light Requirement: Sun
Soil: Moist to medium dry; loam, clay loam
Native Habitat: Prairies, savannas, old fields
Native to: AL AR CT DE IL IN KS KY LA MA MD MN MO MS NC NH NJ NY OH OK PA RI SC TN TX VA WI WV; native but rare in GA NE SD VT
USDA Hardiness Zones: 3–9

I love American Goldfinches and wanted to grow their favorite plant in my yard. So in fall a few years ago, I collected some prairie thistle seed from an old field and simply scattered the seed in a suitable place in my yard—fairly sunny and not too dry. The thistle seeds germinated and prospered beyond my expectations. Every year, they surprise me by popping up here and there in my yard. Even though the plants are usually biennial, some seeds must have taken an extra year to germinate, since I now have plants every year, not just alternate years. Each summer, I wait impatiently for these late bloomers to flower and seed, so I can enjoy goldfinch feeding on the seeds and collecting the fluff. An additional joy is the sight of so many visiting insects. Last year, I was thrilled to see three great spangled fritillary butterflies nectaring on my prairie thistles on a single day in late August!

Swamp or Fen Thistle (*Cirsium muticum*)

Biennial Wildflower
Size: 3–9 feet
Color and Season: Rose-purple flowers in summer
Light Requirement: Sun to moderate shade
Soil: Wet to moist; sand, limestone seeps
Native Habitat: Moist prairies, wetlands, fens
Native to: CT IL IN KY NC ND NH NJ NY MA ME MI MS MN MO OH PA RI TN VA WI WV; native but rare in AL AR FL GA IA LA MD OK SC VT TX; extirpated in DE
USDA Hardiness Zones: 3–9

The swamp thistle has vibrant rose-purple flowers with a cobwebby base, which is sticky to the touch.
PHOTO BY ANNE MORETTI

I enjoy watching for this thistle—one of my favorites—in wetlands in mid to late summer. It has vibrant rose-purple blossoms. The base of the flower heads is cobwebby, nearly spineless and sticky to touch. This stickiness is one of the easiest ways to identify this species. I never fail to see bees and butterflies visiting the flowers for nectar.

PURPLE CONEFLOWER (*Echinacea* species)

Aster or Composite Family, Asteraceae or Compositae

Purple coneflowers are prairie classics and among the most popular native garden flowers. They have showy purple to lavender reflexed petals surrounding purplish cone-shaped disks with orange-tipped spines. The genus name, *Echinacea*, comes from a Greek word for *hedgehog*, an allusion to the spiny disk. There are ten species found only in the eastern and central United States. Two of the more commonly grown species are featured here: pale purple coneflower (*E. pallida*) and purple coneflower (*E. purpurea*). The easiest way to distinguish

Purple coneflowers are favorites of hummingbirds and gardeners alike.
PHOTO BY DAWN-MARIE STACCIA

the two species is by their leaves. The pale purple coneflower has long narrow leaves, while the purple coneflower has broader leaves.

Value to Birds and Other Wildlife

Hummingbirds sip nectar from the flowers and snack on the tiny insects that visit for nectar and pollen. Finches also seek out the insects for their nestlings. Later, in late summer and early fall, goldfinches devour the nutritious seeds, and in winter, juncos come to join them. *Echinacea* flowers are magnets for butterflies. Twenty-two species have been documented nectaring on the flowers, including the red admiral, Baltimore checkerspot, painted lady, American lady, monarch, clouded sulphur, viceroy, fritillaries, skippers, swallowtails, and others. Caterpillars of the silvery checkerspot feed on the foliage, while several moth caterpillars feed on the flower heads. The newly hatched silvery checkerspots winter in the leaf litter, so be sure to minimize or avoid winter and spring cleanup. Many bees, especially long-tongued bees like the bumblebees, visit the flowers for both nectar and pollen. The mining bee (*Andrena helianthiformis*) specializes in this species.

Among mammals, deer may eat the plants but cause few problems. Moles and gophers can damage the roots.

How to Grow

Both pale purple and purple coneflowers are wonderful plants for prairie gardens and perennial borders. They are easy to grow in sunny locations and differ

Purple coneflowers have nectar-rich flowers that attract monarchs and many other butterflies.
PHOTO BY DAWN-MARIE STACCIA

mainly in their moisture needs. The pale purple coneflower prefers drier soils, while purple coneflower likes moist soils. The pale purple coneflower blooms a month earlier than the purple coneflowers. Plant both species for a continuous, months-long floral display of these beautiful blooms.

Choose straight native species of purple coneflowers. They are beautiful and hardy plants. Avoid cultivars and hybrids, which can more easily succumb to extreme weather and disease because they've lost a portion of the rich gene pool of the straight native species that enables them to better adapt to such stresses. For example, Neil Diboll of Prairie Nursery in Wisconsin reported the loss of as many as 90 percent or more *Echinacea* hybrids during the polar vortex in the winter of 2014. Many *Echinacea* cultivars have been removed from botanical gardens due to such problems. Straight native *Echinacea* are also more likely to provide nectar and pollen for insects, since some cultivars are sterile.

Propagation

Seeds; dividing not recommended because divisions result in fewer flowers.

Companion Plants

Plant in clusters along with other prairie flowers and grasses, including those highlighted in this chapter. Yellow-flowered species like yellow coneflower (*Ratibida pinnata*), stiff coreopsis (*Coreopsis palmata*), and black-eyed Susan (*Rudbeckia hirta*) offer a lovely contrast to the purples and lavenders of the *Echinacea* flowers.

A Closer Look

Take the time to watch the unfolding of these beautiful blossoms. The ray flowers, surrounding the disc, open first. Then the tiny disc flowers open gradually from the outer edge of the disc to its center. Look closely at the disk and you may be lucky to detect the caterpillars of the wavy-lined emerald moth, also called the camouflaged looper. It covers its body with the violet-hued stigmas of the flowers of pale purple coneflower. Check other plants as well because this master of disguise takes on the colors of any flowers it visits, including black-eyed Susans (*Rudbeckia* species) and blazing stars (*Liatris* species), among others.

Later in the season, enjoy the many goldfinches that perch on the dark seed heads to feed, seemingly weightless golden gems, rarely bending the sturdy stems, fluttering about now and then, bringing life and beauty to the prairies.

Pale Purple Coneflower (*Echinacea pallida*)

Perennial Wildflower
Size: 3–5 feet
Color and Season: Pink flowers in summer
Light Requirement: Sun
Soil: Dry to medium dry; clay, loam, sand
Native Habitat: Prairies
Native Distribution: AR IA IL IN KS LA MO OK TX; native but rare in AL MN NC NE TN WI
USDA Hardiness Zones: 4–8

A prairie icon along with its cousin purple coneflower (*Echinacea purpurea*). The lovely pale pink flowers blossom for three weeks in early summer, when few other sun-loving plants are in flower. This species is very adaptable and tolerant of drought, heat, and humidity and grows well in poor soils, which, no doubt, is why it flourishes in the gravelly septic field in my yard, where I have a small prairie. The flowers tend to flop if overwatered or lacking in support from other close-by prairie flowers and grasses.

The graceful pale purple coneflower is tolerant of drought, heat, humidity, and poor soils.
PHOTO BY MARIETTE NOWAK

Purple Coneflower (*Echinacea purpurea*)

Perennial Wildflower
Size: 3–4 feet
Color and Season: Purple, rarely white, flowers in summer
Light Requirement: Sun, partial shade
Soil: Dry, medium; clay, loam sand
Propagation: Easy from seed
Native Habitat: Savannas, open woods
Native Distribution: AL AR IL IN KY MO OH OK TN; native but rare in IA FL GA KS LA MS NC SC TX
USDA Hardiness Zones: 4–8

Beloved by butterflies and finches, this beauty blooms profusely for up to six, even eight weeks in midsummer and sometimes reblooms in fall. It prefers moist soils and is the least drought tolerant of the *Echinacea* species. Occasionally, it can become aggressive in perfect conditions. The flowers are long-lasting in the vase, and the seed heads can add a bold statement in dried floral arrangements.

SUNFLOWER (*Helianthus* species)

Aster or Composite Family, Asteraceae or Compositae

Our native sunflowers are gorgeous and rambunctious wildflowers with abundant, bright yellow blossoms. More than three dozen species are native to North America. All are perennial, except for the huge annual common sunflower (*Helianthus annuus*), a mostly western species introduced throughout the United States. Habitats vary, but all prefer sun to part sun. Most grow four to six feet tall, but the common sunflower can reach ten feet.

Many sunflower species are aggressive, spreading by rhizomes. They also have toxic allelopathic compounds in their tissues, which can wash onto nearby plants and soils, harming other plants and even affecting the sunflowers themselves in time. Nonetheless, sunflowers are important prairie plants, invaluable to birds and other wildlife, and should be included in most sunny landscapes.

I grow two sunflower species in my prairie. The western or few-leaved sunflower (*H. occidentalis*) is one of the shortest and least aggressive native sunflowers, making it an ideal choice for pocket prairies and gardens. The prairie

Sunflowers, like this western sunflower, have large, nutritious seeds for songbirds.

PHOTO BY PETER M. DZIUK, www.MinnesotaWildflowers.info

sunflower (*H. pauciflorus*), with its two-tone flower heads, is a favorite of mine, although it can be aggressive. Most of the other species are far too big for small gardens but would be fabulous additions to larger prairie landscapes.

A note of caution: Jerusalem artichoke (*H. tuberosus*) has edible tubers but is considered "more aggressive than Genghis Khan," according to *New York Times* garden expert Leslie Land, and has no place in a small garden.

Value to Birds and Other Wildlife

Hummingbirds sometimes sip nectar from the flowers. A multitude of bees, butterflies, and other beneficial insects, likewise, are attracted by the abundant nectar. Some bees also collect pollen. Other insects feed on the leaves, stems, roots, and other parts of the plants. Sunflowers are one of the premier hosts for lepidoptera; among herbaceous plants, they are third after goldenrods and asters in the number of butterfly and moth caterpillars that feed on the plants and host more than sixty species. Butterfly caterpillars of the checkerspot and the bordered patch feed on the foliage. Several moth caterpillars specialize in sunflowers, including the sunflower borer moth and sunflower bud moth. In addition to caterpillars, many other kinds of insects feed on sunflowers: ants, beetles, bugs, flies, grasshoppers, leafhoppers, and thrips. All these abundant insects offer nutritious food for birds throughout the summer.

Later in the season, the large and nutritious seeds of sunflowers are a favorite food of American Goldfinches, as well as many other songbirds, including the Mourning Dove, American Crows, Black-capped Chickadee, Tufted Titmouse, White-breasted Nuthatch, White-winged Crossbill, and several species of sparrows, meadowlarks, and blackbirds, plus game birds like the Ring-necked Pheasant, Northern Bobwhite, and Sharp-tailed Grouse.

Ground squirrels, chipmunks, gophers, voles, rabbits, deer, and livestock feed on the seeds and other parts of sunflowers.

How to Grow

Sunflowers are hardy and easy to grow when planted in their preferred habitats. The main thing is to keep them in check. In small gardens, use a sharp spade around the plants each spring, and pull out any that wander. Or consider using garden edging to enclose the patch, as my husband did to control our western sunflowers. A third option is to divide the plants every three to four years. This last method is the best way to maintain plant vigor.

Propagation

Easy from seeds or cuttings. Take five to seven node cuttings from top of plant in later spring and plant deep in a pot. Often blooms the same year.

Companion Plants

Almost any sun-loving prairie plants adapted to similar soils are suitable to grow with sunflowers, but they will look particularly striking when grown with lavender-flowered plants like wild bergamot (*Monarda fistulosa*) and blazing stars (*Liatris* species) interspersed with native grasses.

A Closer Look

Throughout the summer, search for the caterpillars feeding on sunflowers, although songbirds may snatch up these nourishing critters first. Also look for galls, spindle-shaped bulges on sunflower stems, which swell up when the gall midge (*Neolasiopterа thurstoni*) lays its eggs on the plant. The gall is a safe home for the hatchlings that feed inside the galls until emerging as adults. Finally, in late summer into fall, don't miss goldfinches and other songbirds swarming stands of sunflowers to gobble up the seeds.

Prairie or Stiff Sunflower
(*Helianthus x laetiflorus; H. pauciflorus; H. rigidus*)

Perennial Wildflower
Size: 3–6 feet
Color and Season: Yellow flowers in summer
Light Requirement: Sun
Soil: Dry to medium; loam, clay loam, sand
Native Habitat: Prairies
Native to: CO IA ID IL KS MN MI MO MT ND NE NM OK SD TX WI WY; rare in AR
USDA Hardiness Zones: 3–8

This gorgeous sunflower, formerly called showy sunflower for good reason, has blossoms up to three inches wide with reddish-purple central disks that set them apart from most other yellow-centered sunflowers. It can become quite aggressive, but this can be useful in some situations. For example, I planted a few prairie sunflowers low on a small, dry hill in my yard. Over the years, the sunflowers slowly crept their way up and wiped out the unwanted non-native plants, with no effort on my part. Soon, however, they also invaded adjacent native plantings, spreading widely by rhizomes. Moreover, they themselves became overcrowded, producing only a few blossoms a year. Although wonderful plants for large expanses, this species is not recommended for small gardens.

Western Sunflower (*Helianthus occidentalis*)

Perennial Wildflower
Size: 2–5 feet
Color and Season: Yellow flowers in summer
Light Requirement: Sun
Soil: Dry, medium; loam, sand
Native Habitat: Prairies, barrens
Native to: AL AR FL IA IL IN KY MI MN MO OH WI; native but rare in GA KS LA MD MS SC TN TX VA WV; extirpated in NC PA
USDA Hardiness Zones: 3–9

Sporting brilliant, starlike yellow flower heads, this species is easy to identify because, unlike other sunflower species, it has only a pair or two of leaves on an otherwise naked flower stalk, giving it an alternate name, naked stemmed sunflower. Although it spreads by rhizomes like most other sunflowers, it is one of the less aggressive sunflower species. This, along with its modest height compared to other sunflowers, makes it an excellent choice for gardens and pocket prairies. Its common name, as well as its species name, *occidentalis*, refers to the West, even though it is native no farther west than eastern Texas and Minnesota. This misnomer came about because it was named by an eighteenth-century botanist when Pennsylvania was considered "the West." Sadly, it has now been extirpated from that very state.

SWITCHGRASS (*Panicum virgatum*)

Grass Family, Poaceae
Perennial Grass
Size: Height 3–6 feet, width 2–3 feet
Color and Season: Purple stigmas at flowering in summer; golden fall color
Light Requirement: Sun
Soil: Moist to dry; clay, loam sand
Native Habitat: Prairies, savannas, damp roadsides
Native to: All lower 48 states except CA, OR, WA; native but rare in ID VT
USDA Hardiness Zones: 3–10

A widespread grass found throughout the original North American prairies, this handsome species is now widely used in native landscaping. Its profuse airy panicles, seven to twenty inches long, dangle with tiny reddish-purple flowers in late summer that mature into multitudes of golden-brown seeds. The green foliage becomes golden late in the year, adding a splash of color to the winter landscape. It is very adaptable to a wide range of conditions, tolerating

When switchgrass seeds are ripe, they fall to the ground, where they're eaten by many ground-feeding songbirds and game birds.
PHOTO BY MARIETTE NOWAK

poor drainage and occasional flooding. Research has been done on using this grass as a biofuel.

Value to Birds and Other Wildlife

This grass is an outstanding source of seeds for many ground-feeding songbirds and game birds—Sora, snipe, quail, turkey, Prairie Chicken, pheasant, Mourning Dove, woodcock, Horned Lark, pipit, Pectoral Sandpiper, Red-winged Blackbird, cowbird, Bobolink, Dickcissel, cardinal, grosbeak, cowbird, Brown Creeper, Pine Warbler, junco, longspurs, and fifteen sparrow species. Ducks and Canada Geese eat the seeds, and some are known to also feed on fresh foliage. The grass is also used for nesting material. The dense foliage remains upright much longer than big bluestem and Indian grass and offers excellent cover for birds and small mammals in winter. More than twenty species of caterpillars hosted by switchgrass provide nourishing food for adult and nestling birds alike.

Many skipper caterpillars, including the widespread Delaware and tawny-edged skippers, feed on switchgrass. The grass is also an essential larval host for most satyrs, like the common wood nymph and the little wood satyr.

Mice, rabbits, and muskrats eat the foliage. It is also often planted for forage for cattle.

How to Grow

Long a favorite in European gardens, this beautiful, adaptable, and easy-to-grow tall grass is now popular here in its native land too. Switchgrass has many versatile uses in the garden—as a screen, specimen or border plant, slope and stream bank stabilizer, container plant, and an excellent species for pocket prairies as well as large prairie restorations. In preferred conditions, it can be aggressive and may need to be controlled in very small landscapes; at the same time, this makes it a great choice to compete with invasive species. It is also tolerant of salt and can do well in roadside ditches.

Propagation

Seeds; division in spring.

Companion Plants

Plant with mid to late summer prairie flowers, such as purple coneflowers (*Echinacea* species), goldenrods (*Solidago* species), and asters (*Symphyotrichum* species), including many of the species in this chapter.

A Closer Look

Enjoy the bouncy flight of the common wood nymph and little wood satyr, both of which feed on grasses in prairies and open fields. If you have good eyes, try to search for tiny ants at the base of switchgrass clumps in moist prairies or old fields. The ant species (*Lasius umbratus*) is at most one-fifth of an inch (5.5 mm) in size. The worker ants, which you are more likely to see, are yellowish brown; the males and queen are medium to dark brown. They nest among the roots of the switchgrass. Amazingly, an even smaller black or brown ant (*Ponera pennsylvanica*)—no more than one-eighth inch in size—is also found within the *Lasius* nests. Both species are common in most of eastern North America.

Later in the season, the ripe grass seeds fall to the earth, providing food for many ground-feeding birds. Not so easily seen, the birds feed among the thicket of lower leaves often hidden from view. But it's worth taking a quiet time, listening and looking for the many sparrows and other birds that you may observe searching for seeds beneath the stems and foliage.

YELLOW CONEFLOWER (*Ratibida pinnata*)

Aster or Composite Family, Asteraceae or Compositae
Perennial Wildflower
Size: 3–6 feet
Color and Season: Yellow flowers in summer and early fall
Light Requirement: Sun to part shade
Soil: Dry to moist; clay, loam, sand
Native Habitat: Prairies, savannas
Native to: AR IA IL IN GA KS KY LA MI MN MO MS NE OH OK SD TN WI; native but rare in AL FL PA SC TX
USDA Hardiness Zones: 3–9

Yellow coneflowers are prairie classics with anise-scented seeds.
PHOTO BY DAWN-MARIE STACCIA

Yellow coneflower is a prairie classic with multitudes of striking two-and-a-half-inch composite flowers with gracefully reflexed petals. The central disk is usually gray, giving it the alternate name gray-headed coneflower. The seed heads have a wonderful anise scent when crushed. The blooming season is amazingly long, extending from early summer into early fall. The plants flutter and sway with the breeze, giving them an airy, carefree quality. Both flowers and seed heads are attractive in floral arrangements.

Value to Birds and Other Wildlife

The seeds are a favorite of American Goldfinches and other songbirds and game birds. The plant hosts caterpillars of the silvery checkerspot butterfly and several moth caterpillars, including the wavy-lined emerald and common eupithecia. Many bees, butterflies, flies, small butterflies, and beetles visit for nectar. Some bees also collect the pollen. Soldier beetles feed on both nectar and pollen. The wasp mantidfly preys on the other insects that visit the flowers. Groundhogs and other mammals eat foliage and flowering stems of young plants.

How to Grow

An excellent choice for prairie gardens and perennial borders. Easy to grow, and hardy, it is able to survive heat and drought. Yellow coneflower is an early succession plant in prairie gardens and can be crowded out by later-emerging, more dominant species. To maintain it, I weed out small areas around the coneflowers to give them enough space to thrive.

Propagation

Easy from seeds. Self-sows readily and seedlings can be transplanted.

Companion Plants

The soft yellow color of these coneflowers looks lovely with other summer-blooming plants that have contrasting colors like blazing stars (*Liatris* species), butterfly weed (*Asclepias tuberosa*), prairie clover (*Dalea purpurea*), and the greens stems of prairie grasses.

A Closer Look

Watch for the many insects that visit yellow coneflower blossoms. Black and gold bumblebees (*Bombus auricomus*), one of the largest species of bumblebees, perch on the flower heads, all facing the same direction watching for other visiting males and females. When rival males arrive, the bees chase them away and then quickly return to their perches. Also search the disks for the tiny moth caterpillars of the wavy-lined emerald (*Synchlora aerata*) and the common pug (*Epithcesia miserulata*). The emerald caterpillar, about a quarter inch long, disguises itself with the disk florets and other plant parts. The pug, up to about a half inch long, has variable colors, often matching the plant on which it feeds, but doesn't camouflage itself with plant parts. Both are in the Geometridae or inchworm family, so you may see them "inching along" when they move. If you're lucky, you may also discover the caterpillars of the silvery checkerspot butterfly feeding on the leaves. They are black with a white dusting and may have a yellow-orange stripe or two on the side.

As the flowers fade, the goldfinches are the show of the season. They perch on the slim stems, which, unlike those of purple coneflowers, bend under their weight as they feed. There can be dozens of finches feeding voraciously in a stand of yellow coneflowers, providing a lively scene in late season prairies.

SILPHIUMS (*Silphium* species)

Aster or Composite Family, Asteraceae or Compositae

> What a thousand acres of Silphiums looked like when they tickled the bellies of the buffalo is a question never again to be answered, and perhaps not even asked.
>
> —Aldo Leopold, *A Sand County Almanac*

Silphium species are the stars of the prairie. Dramatic towering plants, they have saucer-sized, bright yellow flowers and distinctive, bold, sandpaper-like leaves. Whether *Silphiums* are scattered plants standing tall like sentinels or in lush, dense colonies amid a sea of grass, they truly are a sight to savor. These hardy plants can live for decades due to their amazing roots, which can go down twenty feet or more, enabling them to survive droughts. Renowned conservationist Aldo Leopold described trying to dig one up, writing, "After half an hour, of hot, grimy labor the root was still enlarging like a great vertical sweet-potato." Their tough leaves, which slow the loss of water through transpiration, offer another survival strategy. *Silphiums* can even rebound from damaged roots.

They differ from many other plants in the aster family in that their ray flowers—the individual flowers around the central disk of the flower head—are

American Goldfinches devour cup plant seeds in late summer and early fall.
PHOTO BY JACK BARTHOLMAI

female, while the tiny disk flowers in the center are male. As a result, the seeds of *Silphium* species are produced only around the rim of the flower heads from the ray flowers—in contrast to sunflowers (*Helianthus* species), in which only the central disk flowers produce seeds.

The four main "*Silphium* sisters" to be highlighted are compass plant, cup plant, prairie dock, and rosinweed. It is easy to distinguish them by their leaves. Prairie dock has wide, basal, heart-shaped leaves, while compass plant's basal leaves are deeply dissected. The other two have leaves opposite one another along the stem. Rosinweed's opposite leaves rotate their direction 90 degrees as they ascend the stem. Cup plant has leaves that join together around the stem to form a cup, which can hold water temporarily after rains.

Value to Birds and Other Wildlife

Songbirds, especially goldfinches and Sharp-tailed Grouse, devour the seeds in late summer and fall. Meadowlarks, Henslow's Sparrows, Dickcissels, and other prairie birds like to perch on the tall stalks when singing. Goldfinches find shelter in dense colonies of cup plant during the heat of the day where they rest and search for insects. They also drink from the water retained in the cups of this species, as do butterflies. Goldfinches have even been known to take an occasional bath in the cups. Hummingbirds nectar on the flowers.

Insects abound in *Silphiums*. Long-tongued bees, butterflies, and skippers visit the flowers and are the primary pollinators.

A tiger swallowtail sips nectar from the flower of a cup plant.
PHOTO BY DAWN-MARIE STACCIA

Butterflies attracted to the flowers include the silvery checkerspot, bronze copper, red-spotted purple, question mark, monarch, fiery skipper, swallowtail, viceroy, and checkered white. Many bumblebees, miner bees, large leaf-cutting bees, and others visit the flowers. The many other insects feeding on *Silphiums* include weevils, bugs, aphids, ants, beetles, and wasps.

Cattle devour *Silphium* leaves, stems, and flowers. In *A Sand County Almanac*, Aldo Leopold reported that cows in a virgin prairie "cropped the *Silphium* to the ground before any other plant was visibly eaten at all." Bison likely did so as well, but they roamed the prairie and were not restricted to one area where they might devour all the *Silphium* like cattle. Rabbits avoid *Silphium* because of their rough, sandpaper-like leaves.

How to Grow

Silphiums are easy to grow and will flower within two to three years from seed. They are very drought tolerant and thrive in a wide range of soil moistures, with the exception of cup plant, which prefers wetter sites. Amazing as they are, *Silphiums* are not generally recommended on small sites, where they can overwhelm many other desirable species. Nonetheless, we grew several compass plants as a backdrop to our small prairie in front of our former home without a problem, simply cutting newly emerging stems as needed. As for cup plant, its aggressiveness can be advantageous in wetlands, where it competes well with reed canary grass and other invasive plants.

Propagation

Slow but easy from seeds, or transplants. Avoid trying to divide because of their deep and extensive roots.

Companion Plants

Many prairie species are common and beautiful associates with *Silphiums*, including sunflowers (*Helianthus* species), coreopsis (*Coreopsis* species), coneflowers (*Echinacea* species), and blazing stars (*Liatris* species), as well as prairie grasses like Indian grass (*Sorghastrum nutans*), big bluestem (*Andropogon gerardii*), and switchgrass (*Panicum virgatum*).

You never know what you'll find. The author discovered this pearly wood-nymph moth resting on a rosinweed leaf, even though its caterpillars feed on other native plants.
PHOTO BY MARIETTE NOWAK

A Closer Look

Watch for goldfinches, hummingbirds, and other birds that visit *Silphiums* throughout the spring and summer to sip nectar, snag insects, or sing from their tall stems. And bird lovers should be sure not to miss the opportunity to visit a dense stand of these plants later in the season when their seeds ripen. The seeds are as big as the sunflower seeds used in winter bird feeders, and they are just as enticing for birds. I have many fond memories of the goldfinch heaven that was the dense colony of cup plants in the lower prairie at the nature center where I worked. The finches were everywhere gobbling up the seed, chattering away, and seemingly enjoying the glorious fall days as much as I did—a glorious finale to the year before bleak November days arrived.

Rosinweed (*Silphium integrifolium*)

Perennial Wildflower
Size: 3–6 feet
Color and Season: Yellow in summer
Light Requirement: Sun
Soil: Dry to medium; clay, clay-loam
Native Habitat: Prairies
Native to: AR IA IL IN KS KY LA MO MS NE OK SD TN WI; native but rare in IN CO MI SD TX WY
USDA Hardiness Zones: 4–8

Although less dramatic in appearance than some of its gigantic sister *Silphiums*, this species is an ideal choice for smaller prairie gardens because it is shorter, matures faster, and tolerates drought as well as or better than the others. A gorgeous plant, with a multitude of two- to three-inch-wide blossoms and buds, it has a long blooming period of one to two months. It spreads slowly to form small colonies, as it has done in my yard, but it can become aggressive in ideal conditions.

Rosinweed has a resinous sap that Native Americans used as chewing gum.
PHOTO BY MARIETTE NOWAK

Compass Plant (*Silphium laciniatum*)

Perennial Wildflower
Size: 3–10 feet
Color and Season: Yellow flowers in summer
Light Requirement: Sun
Soil: Dry, medium dry; clay, loam, sand
Native Habitat: Prairies
Native Distribution: AR IA IL IN KS LA MN MO MS NE OK SD TX WI; native but rare in KY MI OH SD TN; extirpated in CO
USDA Hardiness Zones: 4–9

Compass plant has a tall flowering stem with up to thirty eye-catching blossoms, each three to four inches across, plus huge dissected basal leaves. Its leaves tend to align in the north–south direction, especially at midday to avoid facing the sun's hot rays, and pioneers were said to use this "compass" for directions. I find that's not always reliable, however, as the leaves are not always aligned strictly north–south. A more likely use of the plant was reported by John Madson in *Where the Sky Began*: "Pioneers sometimes used compass-plant stalks to mark the edges of wagon routes over the wild prairies—tying scraps of cloth to the tall stems to indicate safe passage around boggy swales and sloughs."

Compass plants stand like sentinels in the prairie with bright yellow blossoms and dissected leaves.
PHOTO BY MARIETTE NOWAK

Cup Plant (*Silphium perfoliatum*)

Perennial Wildflower
Size: 3–12 feet
Color and Season: Yellow flowers in summer
Light Requirement: Sun, part sun
Soil: Medium, moist, wet; clay, loam, sand
Native Habitat: Prairies, wetlands
Native to: AL AR IA IL IN KS KY MN MO MS ND NE OH OK PA SD TN TX VA WI WV; native but rare in LA MI NC
USDA Hardiness Zones: 3–8

This bold, beautiful plant has been touted as the single best species for attracting birds. With its cupped leaves that hold water after rains and its nourishing seeds, it is a combo water source and bird feeder. It has a long blooming season, with ten to thirty showy, bright yellow flower heads on a square stem. The species name *perfoliatum* comes from the look of its clasping leaves, as if they've been perforated or pierced by the stem. It's the most aggressive of the four *Silphiums* and spreads readily, although in poor soils like mine it has not done so. Because of its vigor, it competes well with invasive wetland plants like reed canary grass.

Prairie Dock (*Silphium terebinthinaceum*)

Perennial Wildflower
Size: 3–10 feet
Color and Season: Yellow flowers in summer
Light Requirement: Sun
Soil: Medium, moist; clay, loam, sand
Native Habitat: Prairies
Native to: AL AR GA IA IL IN KY MI MO MS NC NY OH SC TN VA WI WV
USDA Hardiness Zones: 4–7

This magnificent prairie classic has clusters of huge leaves at its base that mature in early summer. The flowering stalks shoot up later in summer with big, bold flowers, two to three inches wide, that bloom for a month or longer.

It competes well with the taller, often aggressive, prairie grasses like big bluestem (*Andropogon gerardii*) and Indian grass (*Sorghastrum nutans*). Tough and very drought tolerant, prairie dock will live for decades, once established.

GOLDENROD (*Solidago* and *Oligoneuron* species)

Aster or Composite Family, Asteraceae or Compositae

There are more than a hundred goldenrod species, mostly native to North America, including Mexico. The handsome golden flowers are beloved by many. British gardeners were among the first to prize American goldenrods as garden plants, long before we did. Eventually, we've grown to appreciate them as well, and they've been chosen as the state flowers of Kentucky, Nebraska, and South Carolina. Now they are widely promoted by both horticulturalists and native landscapers.

Every sunny landscape, whether a pocket prairie garden or a sprawling grassland, should have one or more goldenrod species. A gorgeous golden cluster of goldenrods in your yard or a dazzling sweep of the masses of goldenrods across an expansive prairie in the late summer sun is a sight not to be missed. Goldenrods not only are beautiful but are vital plants for birds and other wildlife as well.

This stiff goldenrod, like other goldenrods, supports a wide diversity of bees and other insects.
PHOTO BY MARIETTE NOWAK

A common myth is that goldenrods cause hay fever, which is *not* true. Their showy flowers evolved to attract insect pollinators. Their pollen is sticky and too heavy to be carried by the wind. Rather, insects carry the pollen from one goldenrod to another, ensuring cross-pollination. In contrast, the small, light, wind-blown pollen grains of ragweed's inconspicuous flowers are the true culprits of late summer allergies.

The goldenrod species featured in this chapter—old field goldenrod (*Solidago nemoralis*), showy goldenrod (*S. speciosa*), and stiff goldenrod (*Oligoneuron rigidum*)—are adapted to sunny, open meadows and prairies and will also thrive in sunnier spots in savannas. Other goldenrods are specific to savannas and woodlands.

A female American Goldfinch feeds on the seeds of goldenrod.
PHOTO BY JACK BARTHOLMAI

Value to Birds and Other Wildlife

Goldenrods are worth their weight in gold, serving as islands of life for an amazing variety of birds and insects. Black-capped Chickadees and Downy Woodpeckers dig into goldenrod galls for the larvae sheltering within. (Galls are swellings on plants caused by insect larvae, which create a safe place for them to feed and mature.) Goldfinches, juncos, and sparrows eat the seeds. Goldenrod seeds, flower heads, and leaves are eaten by the Greater Prairie Chicken, plus Ruffed, Sharp-tailed, and Spruce Grouse.

Goldenrods' greatest value for birds, however, is the wonderful diversity of insects that feed on the pollen and nectar, foliage, and seeds. All the insects provide vital protein for birds. More than one hundred species of moths and butterfly caterpillars feed and shelter on goldenrods and are especially important, since fat juicy caterpillars are the preferred food for the nestlings of most land birds. Goldenrods are also one of the most important plants supplying nectar to monarch butterflies to help fuel their long journey south in fall. Other insects sustained by goldenrods include eleven species of bees, some of which specialize in goldenrods, as well as ants, beetles, true bugs, gall-making insects, spiders, aphids, grasshoppers, leafhoppers, midges, and walking sticks. Rabbits, deer, voles, and other mammals browse on goldenrods, especially young plants.

How to Grow

Given the wonderful variety of goldenrods, there is a species for every soil and situation. All are very easy to grow and beautiful additions to native landscapes, as long as you avoid Canada goldenrod (*S. canadensis*) and the similar tall goldenrod (*S. altissimum*), both of which are extremely aggressive and will outcompete all other plants. Grass-leaved goldenrods (*Euthamia* species), not considered true goldenrods, can also be aggressive in moist sites, and you may want to avoid them if you have a small garden, although on my dry soils they are very well-behaved.

Propagation

Easy from seeds. Can divide in early spring. Cuttings of four to six nodes on stems in late spring root easily.

Companion Plants

The gilded blossoms of goldenrods mix beautifully with the blue and purple blossoms of late summer flowers like asters (*Symphyotichum* species), gentians (*Gentiana* species), and blazing stars (*Liatris* species).

A Closer Look

Enjoy the beauty of goldenrods, whether en masse or a few blossoms glowing amid other prairie flowers and grasses. Look closely for some of the many caterpillars—usually well camouflaged and hard to detect—that like to munch goldenrod leaves. Watch for monarchs who seek out goldenrods for nectar before migrating. And search for some of the many other fascinating insects partial to goldenrods. The goldenrod soldier beetle, gold with two dark spots on its back, is commonly seen on goldenrods and is easy to identify. Galls containing insects are also fairly easy to spot. The black blister gall (*Asteromyia carbonifera*) looks like a flat white or black blister on the leaves and is created by a midge that lays its eggs on the goldenrods. At first the blister is whitish; later the blister turns black, revealing the fungus underneath, which is all that remains of the gall. The fungus helps protect the growing midge inside the gall from parasitic wasps. I've sliced the blisters in half and found the tiny yellowish-brown grubs inside. Another gall in the shape of a ball on goldenrod stems attracts woodpeckers and chickadees who dig in to find the fly grub inside (as did ice fishermen in the past).

As the season wanes and the goldenrod seeds ripen, it's time to enjoy the many finches, juncos, and sparrows that visit to dine, balancing on the stems with acrobatic agility to reach the seed heads. At the slightest disturbance, the birds dive down to hide among the dense stems, then bounce back when calm returns. The tufted seeds are soon blown away by breezes, eventually settling on the ground, and the birds follow to dine on the ground-level table of the prairies and meadows.

Old Field or Gray Goldenrod (*Solidago nemoralis*)

Perennial Wildflower
Size: 1–2.5 feet
Color and Season: Yellow in late summer, early fall
Light Requirement: Sun, partial sun
Soil: Medium dry, dry; poor soil, even rocky clay slopes
Native Habitat: Prairies, savannas
Native to: All lower 48 states except 7 western states: AZ CA ID NV OR UT WA
USDA Hardiness Zones: 3–9

One of the shortest species of goldenrod, this species can be grown as a ground cover and is especially lovely amid the lavender hues of purple love grass (*Eragrostis spectabilis*) and rough blazing star (*Liatris aspera*) and other hardy, late summer species that also do well in dry, impoverished soils. A carefree plant, it is one of the last goldenrods to flower. It has narrow wand-like inflorescences that arch gracefully to one side, with a blooming period of a month or so. The dense, whitish hairs on its leaves and stems are a distinguishing feature of this goldenrod, giving it the alternate name gray goldenrod.

One of the last goldenrods to blossom, old field goldenrod has arching wands of blossoms.
PHOTO BY MARIETTE NOWAK

Stiff Goldenrod (*Oligoneuron rigidum*; *Solidago rigida*)

Perennial Wildflower
Size: 3–5 feet
Color and Season: Yellow flowers in late summer to early fall
Light Requirement: Sun
Soil: Dry, medium; clay, loam, sand
Native Habitat: Prairies
Native to: AL AR CO IA IL IN KS KY LA MI MN MO MS MT NC ND NE NM OH OK SD TN TX WI WY; native but rare in CT DE GA NC NJ NY MD PA RI SC VA WV; extirpated in MA
USDA Hardiness Zones: 3–9

This stately goldenrod is noted for its large velvety leaves, which turn a pleasing shade of dusty rose in the fall. Its spectacular golden-yellow, flat-topped clusters of flowers have a mild floral fragrance. A hardy, adaptable species, it thrives in a wide variety of soils, from clay to sand. Because it self-seeds readily, the removal of seed heads may be needed in small gardens to prevent spreading (but this will eliminate a seed source for birds).

Showy Goldenrod (*Solidago speciosa*)

Perennial Wildflower
Size: 1–4 feet
Color and Season: Yellow flowers in late summer to early fall
Light Requirement: Sun; partial sun
Soil: Medium to dry; loam, sand
Native Habitat: Prairies, savannas
Native to: AL AR CT IA IL IN GA KS KY LA MA ME MI MN MO MS NC ND NE NY OH OK RI SC SD TN TX VI WI WV; native but rare in DE MD NH NJ PA
USDA Hardiness Zones: 3–8

One of the most beautiful goldenrod species and a favorite of mine, showy goldenrod has dense pyramidal plumes of flowers on attractive, leafy, reddish

Showy goldenrods with their narrow plumes of flowers are attractive goldenrods for the garden.

PHOTO BY MARIETTE NOWAK

stems. This species is especially showy when accented with other late-blooming plants, like rough blazing star (*Liatris aspera*) and sweet black-eyed Susan (*Rudbeckia subtomentosa*). An excellent choice for gardens as well as restorations.

INDIAN GRASS (*Sorghastrum nutans*)

Grass Family, Poaceae
Perennial Grass
Size: Height: 5–7 feet, width 1–2 feet
Color and Season: Bronzy yellow in flower in late summer; golden brown in autumn
Light Requirement: Sun, light shade
Soil: Dry to moist; clay, loam, sand
Native Habitat: Prairies, savannas, open woodlands
Native to: All lower 48 states except ID NV OR WA; native but rare in ME MT VT
USDA Hardiness Zones: 3–9

The bright yellow anthers of Indian grass hang from the stems like tiny bells.
PHOTO BY MARIETTE NOWAK

This species is, in my view, the most beautiful of the prairie tall grasses. It was a widespread grass in the original "waves of grain" in the North America prairies, later to be replaced by cultivated grains. Blooming in late summer, it has handsome, bronzy-yellow panicles of flowers with showy, bright yellow anthers. In early fall, the grass appears even more stunning in the late afternoon sunlight with its plumes of golden-brown panicles, rich with seeds. Fresh or dried plumes of this grass are lovely additions to floral arrangements.

Value to Birds and Other Wildlife

Upland game birds and songbirds, especially finches and sparrows, eat the seeds, as do small mammals. The tall foliage also provides nesting sites and cover for birds, including Ring-necked Pheasant, Greater Prairie Chicken, Northern Bobwhite, Mourning Dove, Field Sparrow, and others.

Many insects feed on the foliage and stems, particularly grasshoppers, and it is the preferred host plant for several, such as the velvet-striped grasshopper, little pasture grasshopper, and handsome grasshopper. Several leafhoppers and the caterpillars of the pepper-and-salt and Ottoe skippers feed on the foliage. All these insects, in turn, provide essential protein food for birds.

Indian grass offers nutritious forage for bison and elk as well as cattle.

How to Grow

Indian grass is one of the easiest species of tall grass to grow and is very adaptable to a wide range of conditions. It is beautiful in prairie gardens, meadows, border plantings, shallow rain gardens, even roadsides. It is also useful for stabilizing slopes and swales. Because it spreads readily by both seed and rhizomes, it can be aggressive and may not be suitable for small gardens.

Propagation

Seeds; self-sows.

Companion Plants

This grass is beautiful in combination with little bluestem (*Schizachyrium scoparium*) and prairie flowers, such as spotted joe-pye weed (*Eutrochium maculatum*), smooth blue aster (*Symphyotrichum laeve*), and wild bergamot (*Monarda fistulosa*).

Indian grass turns a beautiful warm bronzy color in fall.

PHOTO BY MARIETTE NOWAK

A Closer Look

Enjoy the beauty of Indian grass in all seasons. Be sure to examine the bright yellow anthers when the grass is in flower in late summer. And don't miss its amazing bronzy color on a sunny afternoon in autumn! It is then time, too, to keep an eye out for sparrows and finches searching for fallen seeds below the multitudes of stems.

PRAIRIE DROPSEED (*Sporobolus heterolepis*)

Grass Family, Poaceae
Perennial Grass
Size: Height 2–3 feet, width 12–15 inches
Color and Season: Pinkish-brown panicles in summer; golden color in autumn
Light Requirement: Full sun
Soil: Dry to medium moisture; loam, sand
Native Habitat: Prairies, savannas
Native to: AR CO IA IL IN KS MN MO ND NE OK SD TX WI; native but rare in CT GA KY MD MI NC NM NY OH PA TN VA WY; extirpated in MA
USDA Hardiness Zones: 3–8

This is one of the most elegant of prairie grasses, with graceful, arching leaves creating large round tufts. These dense mounds of narrow, glossy leaves are

Prairie dropseed has dense mounds of arching leaves and flowering panicles that smell like popcorn or coriander.
PHOTO BY MARIETTE NOWAK

quite distinctive and resemble no other prairie grasses, making it easy to identify. When flowering in late summer, airy sprays of fragile panicles rise above the leafy mounds and are especially stunning when backlit by the sun. The ripe seeds soon fall to the ground where they are eaten by wildlife, rather than being retained on their standing stems as with many other grasses. Dropseed's foliage stays upright through winter snows, giving it year-round interest. Once common throughout much of eastern North America, this grass is now rare in many states, even designated as endangered in a few. By planting prairie dropseed, gardeners can help restore this beautiful grass to its former range.

Value to Birds and Other Wildlife

Through late summer and winter, the seeds are a favorite of many sparrows, including Field, Savanna, and Tree Sparrows, as well as Slate-colored Juncos (also members of the sparrow family). Bees use the grass for nesting material and also nest within or below the clumps. This grass hosts such rare butterflies as the endangered Poweshiek skipperling and the threatened Dakota skipper, plus a few other lepidoptera species. Several ant species build mounds among the roots and tussocks of this grass. Two rare leafhoppers specialize in feeding on dropseed. Grasshoppers occasionally eat the foliage. Bison, cattle, and horses also will eat the foliage, and small rodents create burrows in the leafy tufts and root mass of this grass.

How to Grow

Although relatively slow growing, once established, this is a trouble-free and long-lived grass at a suitable site. It grows best on moist, rich soil but has deep roots and is very drought resistant.

Propagation

Seeds, division.

Companion Plants

Purple prairie clover (*Dalea purpurea*), black-eyed Susan (*Rudbeckia hirta*), and rough blazing star (*Liatris aspera*) are beautiful native wildflower species that commonly grow with this grass on remnant prairies and are excellent companions in prairie gardens.

A Closer Look

A unique feature of prairie dropseed is the strong fragrance of its flowering panicles. To some, they smell like coriander (also known as cilantro). But to most people, including myself, they evoke the scent of slightly burnt, buttered popcorn. No need to pick the panicles to enjoy the fragrance. Simply squeeze them through your fingers for a stronger smell, as I've taught my grandchildren to do. If you're lucky to have or visit a large stand of the grass, the scent can be enjoyed even from several yards away.

AMERICAN ASTER (*Symphyotrichum* species)

Aster or Composite Family, Asteraceae or Compositae

The name *aster* comes from the Greek word for *star*, and asters are undoubtedly the superstars of the fall landscape. Their beauty is legendary, with colors

Asters are the superstars of the fall landscape.
PHOTO BY SUSAN DAMON

ranging from white and blue to pink and purple. Asters appear to have a single flower, but each flower is actually a composite of many flowers—a center packed with many tiny disk flowers surrounded by colorful ray flowers. The disk flowers are usually yellow at first flowering, but gradually, as they age and are pollinated, the disks become purplish or brown, having lost much of their pollen. The bees take notice and prefer visiting the flower heads with yellow disks, still rich with pollen. So why don't the old flower heads immediately lose their petals and put their energy into forming seeds? It has to do with the value of a larger billboard to catch the eyes of fly-by pollinators. Bees and other insects are more likely to visit a larger floral display that better attracts their attention as they cruise around searching for pollen and nectar. Once landing on the asters, they can simply ignore the spent flowers. This increases the asters' chances for cross-pollination and benefits the insects as well.

Formerly, all asters were in a genus of the same name. But a 1994 study of asters around the world concluded that the American asters should be separated from Eurasian asters based on differences in evolutionary history, size, chromosome shape, and seeds. The Eurasian species kept the genus name *Aster*, as required by nomenclature rules, because they were named first. Most of the American asters, about ninety species in total, were put into the genus *Symphyotrichum*.

The asters featured in this book are all within this genus. There are *Symphyotrichum* asters for woodlands, savannas, and wetlands, as well as prairies. Those highlighted in this chapter are three easy-to-grow favorites for both prairie gardens and large-scale prairie restorations: heath aster (*S. ericoides*), New England aster (*S. novae-angliae*), and smooth blue aster (*S. laeve*).

Value to Birds and Other Wildlife

Songbirds that feed on the seeds in fall and winter include the American Goldfinch, American Tree Sparrow, Black-capped Chickadee, Dark-eyed Junco, House Finch, Pine Siskin, and Song Sparrow. Two game birds, the Wild Turkey and Ruffed Grouse, feed on the foliage as well as seeds. Asters also offer cover for many birds, including waterfowl. One hundred species of butterfly and moth caterpillars feed on aster and, in turn, offer nutritious protein-rich food for adult and nestling birds alike. All the other insects found on asters are also potential food for birds and mammals.

Late-blooming asters offer vitally needed pollen and nectar for bees, butterflies, and other insects in autumn.
PHOTO BY SHEILA McMAHON

Pearl crescent and silvery checkerspot caterpillars are two of the more common butterfly caterpillars that feed on the foliage and flowers of asters. The best-known moth caterpillars that feed on asters are those of the Isabella tiger moth, commonly known as woolly bears. Many adult butterflies and skippers visit asters to nectar on their flowers, including the black swallowtail, common buckeye, eastern tiger swallowtail, great spangled fritillary, monarch, orange sulphur, painted lady, pearl crescent, Peck's skipper, red admiral, and tawny-edged skipper. With their late and long blooming period, asters are particularly vital to migrating monarchs, who depend on the rich nectar to fuel their journey south.

A bevy of other critters can be found on asters, including aphids, ants, bees, beetles, bugs, flies, gall insects, spiders, walking sticks, and wasps. White-tailed deer, cottontail rabbits. woodchucks, and livestock such as cattle, horses, and sheep browse on asters, especially on the young foliage.

How to Grow

Asters are easy-to-grow perennials. The many available species provide choices for any situation and soil moisture. They can be moved in spring or fall and benefit from divisions every three to five years in garden situations, although this is rarely done in restorations. The taller species can get leggy, so prune them back about one-third in late spring for bushier plants. Their lower leaves tend to get mildew and rust by flowering, which can be hidden by planting shorter plants in front of them.

Propagation

Easy from seeds.

Companion Plants

For a last hurrah of beautiful bloom before bleak November sets in, combine asters with late-flowering species like brown-eyed Susan (*Rudbeckia triloba*), bottle gentian (*Gentiana andrewsii*), and sneezeweed (*Helenium autumnale*), interspersed with native prairie grasses.

A Closer Look

Watch for the different ways that birds forage on aster seeds in fall and winter. The American Goldfinch, American Tree Sparrow, Black-capped Chickadee, House Finch, Pine Siskin, and Song Sparrow hang precariously from the plants to glean seeds from the spent flower heads. The American Goldfinch and American Tree Sparrow sometimes use a clever additional technique. While perching atop the plants, they rapidly beat their wings, causing copious amounts of the seed to fall on the ground, where it's easier for them to eat. Dark-eyed Juncos primarily feed on the seeds on the ground, whether blown there by the wind or dropped by the wing-wind of other songbirds.

The heath aster comes to bloom in late summer with dense clusters of bright white-petaled flowers.
PHOTO BY MARIETTE NOWAK

Heath Aster (*Symphyotrichum ericoides; Aster ericoides*)

Perennial Wildflower
Size: 1–3 feet
Color and Season: White flowers in late summer to late fall
Light Requirement: Sun
Soil: Dry to moderate moisture; loam, sand
Native Habitat: Prairies, barrens
Native to: All lower 48 states except CA KY LA NV TX; questionable in NC; native but rare in AL GA MS NJ PA VA; extirpated in TN
USDA Hardiness Zones: 3–9

The many needlelike bracts on the stems of this aster create a heathlike appearance, for which it is named. It thrives in poor sand and gravel soils. Despite the poverty of the soils in which it grows, the plant explodes with hundreds of flowers arranged in dense, showy, often one-sided clusters. It creeps slowly by rhizomes creating patches, as it has done on my small prairie.

New England Aster (*Symphyotrichum novae-angliae; Aster novae-angliae*)

Perennial Wildflower
Size: 3–6 feet
Color and Season: Pink or purple flowers in late summer to late fall
Light Requirement: Sun to partial shade
Soil: Medium, moist; clay, loam, sand
Native Habitat: Prairies, open woodlands
Native to: AL AR CT DE IA IL IN KS KY MA MD ME MI MN MO MS MT ND NE NH NJ NM OH PA RI SC SD TN VA VT WI WV WY
USDA Hardiness Zones: 3–7

Magnificent in blossom, this is one of the most popular and commonly cultivated asters. Each plant has clusters of more than a dozen large, one-inch-wide compound flowers with golden disks surrounded by forty or more ray flowers, ranging in color from violet, lavender, or purple to many shades of pink. The multitude of blossoms and contrasting colors of yellow and purple offer an irresistible display, beckoning myriad pollinators to sample its rich nectar and pollen. Preferring moist soils, this aster is particularly suited for rain gardens, swales, and lakeside plantings but does well in any garden with consistent moisture.

Smooth Blue Aster (*Symphyotrichum laeve; Aster laevis*)

Perennial Wildflower
Size: 2–4 feet
Color and Season: Blue flowers from late summer to October
Light Requirement: Sun
Soil: Dry, medium; clay, loam, sand
Native Habitat: Prairies, woodland edges
Native to: All lower 48 states except CA OR; native but rare in FL GA MS NC SC VT
USDA Hardiness Zones: 4–8

Named for its smooth leaves, smooth aster is a hardy, adaptable plant for the garden.

PHOTO BY PETER M. DZIUK, www.MinnesotaWildflowers.info

A goldfinch feasts on seeds, its favorite food, in the golden glow of sunrise.
PHOTO BY JACK BARTHOLMAI

This aster is named for its long, smooth, blue-green leaves. The similar sky blue aster, in contrast, has rough leaves, as do many other asters. A hardy, adaptable plant, this aster has vibrant starlike flowers that bloom in profusion in open clusters. One of the best asters for gardens, it has a sturdy stem that usually remains upright through the growing season. It can have a long life if not munched down by rabbits or crowded out by more aggressive species, and it persists even after light frosts.

Eastern Bluebirds prefer an open savanna-like habitat with scattered trees and little understory.

PHOTO BY JACK BARTHOLMAI

6 Top 10 Plants for Bluebirds and Other Savanna Birds

Savannas are basically prairies with a few scattered trees, and they once covered vast areas of the Midwest—an estimated 30 million acres. They also occurred in various other areas of the United States. Savannas offer ideal habitat for a great many birds. But sadly today, savannas are among our most endangered plant communities.

The good news is that we can easily create small pockets of savanna habitat right on our own properties. Many homeowners' yards are already similar to a savanna in structure, with one or more trees surrounded by lawn. You can easily transform your yard into a welcoming savanna for birds by replacing some or all of your lawn with native flowers and grasses. Even better, add an oak or several oaks, which are the classic savanna trees in the eastern half of the United States. But nearly any tree, with the exception of invasive species, can suffice as a savanna tree.

There are few plants that occurred solely in savannas. Generally the herbaceous species were a mix of both prairie and woodland species, depending on the available sunlight. Shrubs and small trees occurred in pockets at the edges of savannas or in wetter areas and offered fruit, shelter, and perches for savanna birds. Several of the species highlighted here were very common in remnant savannas, including gray dogwood (*Cornus* racemosa), roses (*Rosa* species), and little bluestem grass (*Schizachyrium scoparium*), even though they occurred in other habitats as well.

The herbaceous wildflowers in this chapter are those that do well in the partial shade typical of savannas and will be beautiful and beneficial replacements for lawn grasses, along with little bluestem grass. Besides oaks, other tree species, such as serviceberries (*Amelanchier* species) and hawthorns (*Crataegus species*), are covered that could be used as alternatives (or additions) to oaks. They would make excellent specimen trees for smaller urban yards and result in mini-savanna settings with the addition of native herbaceous species beneath. Juniper trees and shrubs also occurred in some savannas and are outstanding species for birds and wildlife. All three of these smaller trees—serviceberries, hawthorns, and junipers—have abundant fruit for birds.

This combination of species will host a rich supply of insects and produce bountiful berries for savanna birds. Different birds eat varying amounts of these foods. For example, the Eastern Bluebird's diet is about 70 percent insects and 30 percent berries. Cedar Waxwings, on the other hand, feed almost exclusively on berries, with only about 12 percent of their diet provided by insects. The plants in this chapter will attract bluebirds and waxwings as well as many other common birds found in savannas and nearby shrubby habitats. You may even discover a Red-headed Woodpecker, one of our more rare woodpecker species that has fortunately been increasing where savanna habitat is being restored. Besides the recommended plants in this chapter, there are many other valuable plants that you can gradually add in future years to increase the diversity of your savanna habitat.

SERVICEBERRY (*Amelanchier* species)

Rose Family, Rosaceae

Whether small trees or shrubs, *Amelanchier* species are airy, graceful woody plants that provide year-round beauty in the landscape. Spring brings a plethora of white flowers, summer brings dark blue berries, fall delights with attractive yellow to red foliage, and winter offers smooth, grayish bark—a standout. As a result, *Amelanchiers* are popular landscaping plants, and like many such plants, they have multiple names. *Juneberry* is a name often used because the berries of some ripen in June. The name *serviceberry* or *sarvisberry* comes from the name of a similar European species. Alternately, some say it's because the

Birds of the Savanna

The birds of the savanna habitats are highly variable depending on the density and spacing of the trees, saplings, shrubs, and herbaceous understory. Because of the transitional nature of savanna habitat, savanna birds also tend to overlap with those of prairies and woodlands. As a result, I found considerable variation among ornithologists in their designation of savanna birds. Following is my own compilation of birds typical of savanna habitats based on a number of sources as well as personal experience. I have not included birds that nest and breed primarily in shrubby habitats within savannas, since I've treated them separately in the chapter on shrubland bird habitat.

Eastern Bluebird
Blue Jay
Brown-headed Cowbird
American Crow
Mourning Dove
Northern Flicker
Great Crested Flycatcher
Scissor-tailed Flycatcher
Common Grackle
Sharp-tailed Grouse
Red-tailed Hawk
Swainson's Hawk
American Kestrel
Eastern Kingbird
Western Kingbird
Black-billed Magpie
Baltimore Oriole
Orchard Oriole
American Robin
Loggerhead Shrike
Chipping Sparrow
Eastern Towhee
Black Vulture
Turkey Vulture
Cedar Waxwing
Red-headed Woodpecker

Red-headed Woodpecker populations increase in restored savannas.

PHOTO BY DAWN-MARIE STACCIA

Serviceberry is one of the first trees to blossom in spring.
PHOTO BY ANNE MORETTI

plants bloom in spring when funeral services could be held for pioneers who died in winter when the ground was too frozen for burial. The names *shadbush* or *shadblow* refer to the shad, a fish that spawns in New England rivers when *Amelanchier* is in bloom. *Saskatoon* is a name that originated from a Cree name for a western species. Of historical note, George Washington planted some specimens of *Amelanchier* at his plantation at Mount Vernon.

There are nineteen species in North America with at least one occurring in every state except Hawaii. But it can often be a perplexing genus to identify to species, due to many similar characteristics and multiple naturally occurring hybrids. This complexity has resulted in heated exchanges among taxonomists and misnamed species in nurseries. Two popular species are featured here: the widely planted serviceberry tree (*A. laevis*) and the shrub dwarf serviceberry (*A. spicata*).

Value to Birds and Other Wildlife

Want birds to visit your yard? Then plant one of the shrubs or small trees in this genus—all known to attract a parade of visiting birds. Their berries are a preferred food for at least twenty-five species of birds, including American Goldfinch, American Crow, American Redstart, American Robin, Baltimore Oriole, Bohemian Waxwing, Cedar Waxwing, Brown Thrasher, Eastern Kingbird, Eastern Towhee, Evening Grosbeak, Gray Catbird, Hairy Woodpecker, Hermit Thrush, Northern Cardinal, Northern Flicker, Northern Mockingbird, Rose-breasted Grosbeak, Red-eyed Vireo, Red-headed Woodpecker, Ruffed Grouse, Scarlet Tanager, Swainson's Thrush, Veery, and Wood Thrush. Another thirteen species, as well, occasionally feast on the berries, including the Blue Jay, Common Grackle, Dark-eyed Junco, Downy Woodpecker, Eastern Phoebe, Eastern Bluebird, Mourning Dove, Northern Bobwhite, Red-winged Blackbird, Ring-necked Pheasant, Pine Grosbeak, Song Sparrow, and Tufted Titmouse. My guess is that there are other species yet to be noted that likewise gorge on the berries.

Plenty of insects also find a table at *Amelanchier* species. Of special value for birds are the 124 species of juicy moths and butterfly caterpillars munching on the plant. Among them are those of the red-spotted purple and striped hairstreak butterflies. Bees, beetles, and flies come for the nectar and pollen. Other insects feed on other parts of the plant, including aphids, gall-forming flies, lace bugs, scale insects, saw flies, wood-boring beetles, borers, and weevils.

Wood Thrushes are among the many birds that visit to feed on serviceberries.
PHOTO BY JEREMY MEYER, jeremymeyer.photography

Two insect species, the juneberry lipped gall midge and the serviceberry leaf-roller moth, are named after the genus.

A number of mammals also visit to dine on the berries: bears, chipmunks, foxes, mice, and squirrels. Beavers sometimes gnaw on the bark and wood, while deer and moose browse on the twigs and leaves. Humans, as well, have enjoyed the edible berries since ancient times, and today they are eaten raw or harvested for jellies, pies, and wine.

How to Grow

Serviceberries are extremely easy to grow and adaptable to a variety of soils. Once established, many grow well in poor, droughty soils. They are susceptible to leaf diseases, which can cause leaf fall by early autumn, but this is largely cosmetic and results in no significant damage to the plants. Prune, as needed, to remove dead or crossed branches. The shrub species can be renewed by hard pruning, cutting stems down to three to six inches. Use the tree species as attractive specimen trees or along woodland borders. The shrubs are perfect as ground cover in wilder settings where they can spread with abandon or be used to create barrier hedges.

Propagation

Seeds and cuttings.

Companion Plants

Amelanchier trees stand out beautifully against a backdrop of evergreens. Around their bases, use low-growing herbaceous species that do well in semi-shade, such as elm-leaved goldenrod (*Solidago ulmifolia*), prairie trillium (*Trillium recurvatum*), and Short's aster (*Aster shortii*). Excellent companions for shrub *Amelanchier* species are early low blueberry (*Vaccinium angustifolium*), wild plum (*Prunus americana*), downy arrowwood viburnum (*Viburnum rafinesquianum*), and red osier dogwood (*Cornus sericea*).

A Closer Look

Be sure to enjoy *Amelanchier* year-round. Most species have a very short blooming period, only a week or two, but it's glorious, with every branch adorned with

Some *Amelanchier* berries ripen later than others, which helps extend the berry season for birds.
PHOTO BY MARIETTE NOWAK

delicate blossoms—all the best to attract awakening insects. Don't miss it! After this cloud of blossoms fades, the berries begin to form. Soon it's time for all bird lovers to get their binoculars to watch for some of the nearly forty species of birds that wing their way over for berries, some eaten even before they're ripe. Unlike the flowers that blossom profusely all at the same time, the berries display staggered ripening, with some still green, while others are fully ripe. This extends the berry season, which often lasts from June to August as berries gradually ripen, an adaptation that maximizes the dispersal of the fruit seeds by birds—as well as the time you can enjoy the voracious visiting birds. Next, it's time to enjoy the fall foliage, sporting yellow and red hues. In winter, note the silvery gray bark and branching structure. This is a great time of year to prune, if needed.

Allegheny Serviceberry or Juneberry (*Amelanchier laevis*)

Deciduous Tree or Large Shrub
Size: Height 25–40 feet, width 8–18 feet
Color and Season: White flowers in early spring; orange to dull red foliage in autumn
Light Requirement: Sun, partial shade
Soil: Medium, moist; clay, loam, sand
Native Habitat: Woodlands, wooded ravines, savannas
Native to: AL CT DE GA IA IL IN KY MA MD ME MI MN, NC NH NJ NY OH PA RI SC TN VA VT WI WV
USDA Hardiness Zones: 4–8

I planted this lovely tree decades ago and continue to enjoy it year-round in my yard. The berries are a magnet for a variety of birds and offer a delicious cereal topping for my family. I pick the lower hanging fruit, leaving the upper berries for the birds. The berries are a great substitute for blueberries and contain a plethora of healthy phytochemicals, according to recent research.

A similar species in habitat and adaptability is eastern juneberry, *Amelanchier arborea*. Unfortunately, it has unpalatable fruit and its new leaves are grayish rather than the bronzy purple of Allegheny serviceberry, which sets off its flowers so beautifully in spring. A naturally occurring hybrid between the two species is apple serviceberry (*A. x grandiflora*).

Canadian serviceberry (*A. canadensis*), another tree *Amelanchier*, is native mostly to swamps and low woodlands in Atlantic coast states, but it can be adapted to somewhat dryer sites in the garden.

Dwarf Serviceberry (*Amelanchier spicata; A. stolonifera*)

Deciduous Shrub
Size: Height 1–6 feet, width 3–10 feet
Color and Season: White flowers in spring; variable yellow, orange, or red foliage in autumn
Light Requirement: Sun, partial shade

Soil: Moist, well-drained soil; will tolerate alkaline and clay soils
Native Habitat: Savannas, open woodlands
Native to: CT GA IA IL MA ME MI MN NC ND NH NJ NY OH PA RI SC VA VT WI WV; native but rare in AL KY MD
USDA Hardiness Zones: 4–8

This species is one of a complex of shrubby serviceberries, which have given grief to many a taxonomist trying to delineate precise species, with the result that these shrubs have been given not only many common names but various Latin names as well. They're known as running juneberry, dwarf shadbush, and thicket shadbush, as well as dwarf serviceberry. In many nursery catalogs, one of the most common Latin names for dwarf serviceberry is *Amelanchier stolonifera*, rather than *A. spicata*, the current accepted name. Many similar shrubby serviceberries are native to sandy, rocky soils, but *A. spicata* will adapt to clay as well as alkaline soils. A hardy low-maintenance serviceberry, this is a wonderful choice for open droughty sites, where they'll spread to form rounded mounds or thickets.

GRAY DOGWOOD (*Cornus racemosa*)

Dogwood Family, Cornaceae
Deciduous Shrub
Size: Height 3–8 feet, width 3–8 feet
Color and Season: White flowers in late spring to early summer; purple foliage in autumn
Light Requirement: Sun, partial shade; adaptable to wide range of soils
Soil: Moist to medium dry
Native Habitat: Savannas, thickets
Native to: AR, CT IA IL IN KY MA MD ME MI MN MO ND NH NJ NY OH PA RI VA VT WI WV; native but rare in SC SD NC; extirpated in DE
USDA Hardiness Zones: 4–8

> I would put this on my short list of shrubs important to wildlife.
>
> —William Cullina, *Native Trees, Shrubs, and Vines*

A Gray Catbird devours the last gray dogwood berries.
PHOTO BY JACK BARTHOLMAI

Gray dogwood is one of the most common shrubs found in remnant savannas. It occurred, as well, in nearly all of our original plant communities from prairies to shrub lands, to woodlands and wetlands. This dogwood suckers freely and forms dense thickets on sunny sites, which become havens for birds and other wildlife. But it is the bane of prairie managers, due to this very habit, and for good reason, since prairies and their amazing flora can easily be wiped out by proliferating shrubs. It's essential to preserve all plant communities and their diversity of species, a monumental task for plant managers and, on a smaller scale, gardeners as well.

A hedge of gray dogwood lines part of my driveway, leafy green in early summer, covered soon with panicles of dome-shaped white flowers up to two inches across, and by late summer, sporting clusters of bright white berries on ruby-red stems, and finishing the season with maroon-purple foliage. It's a beautiful shrub for all seasons and a favorite of mine, harboring an amazing diversity of species. But I would not grow it in my prairie.

Value to Birds and Other Wildlife

In summer, gray dogwood provides a wonderful selection of insect food for birds. Most important are the 118 species of caterpillars, rich in protein for birds, that munch on dogwood leaves. Among these caterpillars is the dainty spring/summer azure butterfly, as well as our beloved and beautiful cecropia moth. Three others moth caterpillars are so often found on dogwoods that they are named for the shrub: the dogwood thyatirid, dogwood probole, and dogwood borer moth. Then there are the other countless insects that depend on dogwoods for sustenance. Seven genera of bees, six genera of wasps, plus some species of flies visit for pollen and nectar. Other insects that nibble on parts of dogwoods include aphids, beetles, bugs, midges, and thrips.

Among the many insects are caterpillar-like larvae of the dogwood sawfly. They are fairly common on gray dogwood and feed gregariously on dogwood foliage. After the final molt, they are a striking yellow with black spotlike stripes. In large numbers they can eat most of the leaves, but usually later in summer and not causing permanent damage. As David Smitley of the University of Michigan Extension says, "This is an insect you can enjoy."

As summer wanes, dogwood's branches become laden with berries, which are as irresistible to birds as chocolate is to us. The berries are about a quarter inch in diameter (a perfect size for the beaks of most birds) and very nutritious, containing up to 33 percent lipids, including fats—far more than most wild fruits—and providing important fuel for migrating birds as well as for resident birds. It's no wonder that nearly fifty species of birds have been observed feeding on dogwood berries. Gray dogwood berries are preferred by the American Robin, Cedar Waxwing, Downy Woodpecker, Eastern Bluebird, Eastern Kingbird, Gray Catbird, Northern Bobwhite, Northern Cardinal, Northern Flicker, Pine Grosbeak, Swainson's Thrush, and Wild Turkey.

Other birds feeding on the berries of dogwoods include the American Crow, American Woodcock, Black-capped Chickadee, Blue Jay, Blue-headed Vireo, Brown Thrasher, Crested Flycatcher, Eastern Towhee, *European Starling*, Gray-cheeked Thrush, Greater Prairie Chicken, Hairy Woodpecker, Hermit Thrush, Mallard, Northern Mockingbird, Pileated Woodpecker, Pine Warbler, Red-bellied Woodpecker, Red-eyed Vireo, Red-headed Woodpecker, Ring-necked Pheasant, Rose-breasted Grosbeak, Rusty Blackbird, Scarlet Tanager, Song Sparrow, Summer Tanager, Tree Swallow, Tufted Titmouse, Veery, Warbling Vireo, White-crowned Sparrow, White-throated Sparrow, Wood Duck, Wood Thrush, Willow Flycatcher, and Yellow-bellied Sapsucker. The Ruffed Grouse eats both buds and berries.

The gray dogwood is also a prime nesting site for birds, especially the Gray Catbird. A thicket of these dogwoods also offers cover for many birds.

Not to be left out are mammals, from mice to deer, that eat the berries and seek cover in the dense branches of dogwoods.

How to Grow

This extremely hardy species grows well in moist to dry soils, even sandy and extremely compacted soils. While it tolerates some shade, it flowers and fruits best in full sun. The dense thickets it forms can be controlled and rejuvenated by renewal pruning (cutting stems down to 4–10 inches) every three years or so. Nonetheless, it is difficult to eradicate this shrub in prairies and other sunny locations, so caution should be taken in choosing the site of planting.

Propagation

Seeds and cuttings.

Companion Plants

Grown as a shrub border, gray dogwoods do well with a woodland background of oaks and hickories. The dense hedges themselves don't lend themselves to understory plants, and the shrubs are likely to spread into any herbaceous vegetation planted in the foreground.

Gray dogwood berries are white on bright red stems.
PHOTO BY ANNE MORETTI

A Closer Look

When the berries are ripe, keep an eye out for the many birds that swarm to the shrubs to feed, especially migrants who depend on the fat-rich berries to fuel their long flights south. It's easy to miss them. The shrubs may be full of berries one day and devoid of most berries the next, with the birds long gone before you spot them. Flocks of robins and grackles can feed on as many as fifty berries in one visit, quickly leaving the dogwoods bare!

And don't fail to enjoy the beauty of gray dogwoods even after the birds and berries are gone. The many stems of the berries, now bare, remain brilliant red, framed by the still-green leaves. They are especially striking when seen in the low light of late afternoon in early fall. I once made a bouquet of the showy stems for my table.

HAWTHORN (*Crataegus* species)

Rose Family, Rosaceae

Hawthorns, often called thorn apples, are large shrubs or small trees averaging up to thirty feet tall with thorny branches and small berry-sized fruit. They are native to temperate regions throughout the Northern Hemisphere in Asia, Europe, and North America. Although easy to identify as a genus, they are notoriously difficult to separate into species, with as many as 1,000 species claimed to exist in the past. Today the consensus is that there are about 230 species in the United States of which 17 are hybrids. Every state except Hawaii has at least one native hawthorn, with many having multiple species. Two common hawthorns are featured here: cockspur hawthorn (*Crataegus crus-galli*) and Washington hawthorn (*C. phaenopyrum*). Both are well identified and available at nurseries carrying native trees.

Value to Birds and Other Wildlife

Hummingbirds nectar on the multitude of hawthorn flowers in spring. Many insects are also attracted to the blossoms and are a boon for insect-eating birds. The trees host 159 moth caterpillars and butterfly caterpillars, all providing

Cockspur hawthorn has a showy profusion of flowers in spring.
PHOTO BY KRZYSZTOF ZIARNEK, KENRAIZ, CC BY-SA 4.0, WIKIPEDIA, https://cs.wikipedia.org/wiki/Soubor:Crataegus_crus-galli_kz01.jpg

essential food for nestling birds. By late summer or fall, hawthorns produce loads of berries for birds. Importantly, unlike the fruit of many plants, the berries on many hawthorns persist into winter, providing survival food when little else is available. The twenty species of birds known to feed on hawthorn berries include the American Robin, Blue Jay, Brown Thrasher, Cedar Waxwing, Evening Grosbeak, Fox Sparrow, Gray Catbird, Hermit Thrust, Northern Bobwhite, Northern Cardinal, Northern Flicker, Northern Mockingbird, Pine Grosbeak, Purple Finch, Ring-necked Pheasant, Rose-breasted Grosbeak, Ruffed Grouse, Wild Turkey, Willow Flycatcher, and Wood Duck. The Yellow-bellied Sapsucker dines on the sap of hawthorns. A number of birds are also known to nest in hawthorns, protected from predators by both the dense foliage and thorns, including the American Robin, Blue Jay, Brown Thrasher, Gray Catbird, Northern

Cardinal, Northern Mockingbird, Willow Flycatcher, and Yellow-breasted Chat. Shrikes use the thorns of hawthorns to impale larger insects, small mammals, and birds before either eating them immediately or storing them on the thorns for later consumption.

As noted, insects abound on hawthorns. The nectar and pollen attract small bees, flies, beetles, wasps, and butterflies. Other insects feed on the flowers, foliage, sap, and wood. Among the 159 caterpillar species supported by hawthorns are those of the striped hairstreak and red-spotted purple butterflies, which feed on the foliage. Moth caterpillars include seven underwing species. Other insect feeders include the woolly hawthorn aphid and other aphids, treehoppers, various bugs, beetles, borers, and the hawthorn blossom weevil. Hawthorns are also the preferred host plant for eight species of leafhoppers.

Many mammals eat hawthorn berries, including black bears, coyotes, foxes, rabbits, skunks, and tree squirrels. White-tailed deer and cottontail rabbits browse on hawthorn twigs and foliage.

How to Grow

Hawthorns are very popular landscaping trees, especially for small yards. They should be planted in sunny or mostly sunny sites. Moist soils are preferred, but most can adapt to more droughty areas. Many are susceptible to cedar-apple rust and should not be planted within one hundred yards of juniper shrubs (*Juniperus* species) or red cedar (*J. virginiana*).

Propagation

Seeds and grafts; difficult from cuttings.

Companion Plants

Hawthorns pair well with serviceberries (*Amelanchier* species), gray dogwood (*Cornus racemosa*), and ninebark (*Physocarpus opulifolius*), planted as an informal grouping or hedge.

A Closer Look

Spring and fall are the best times to spot birds in hawthorns. When flowering, hawthorns provide nectar for hummingbirds as well as both nectar and pollen for myriad insects. These insects, in turn, attract warblers whose breathtaking colors are all the better seen against the backdrop of the white flowers of hawthorns. Look for such common warblers as the Chestnut-sided Warbler, Yellow Warbler, and Nashville Warbler. Other bright birds, from cardinals to Indigo Buntings, join in the feast.

From late summer to winter, when the ripe berries become available, don't miss the many birds that visit for nourishment. Migrating birds, like catbirds and Fox Sparrows, fuel up on the berries before departing for southern locales, while resident birds, like cardinals and Blue Jays, fatten up before and during the lean times of winter.

The insects found on hawthorns are of special interest in summer, especially the seven underwing (*Catocala* species) moths that are hosted by hawthorns, one of which is the hawthorn underwing (*C. crataegi*), a specialist that feeds only on hawthorns and apple trees. All seven are found throughout the eastern United States and adjacent central states and most have colorful names, given during the romantic era of the nineteenth century. Besides the hawthorn underwing, they include the charming underwing, wonderful underwing, titan underwing, ultronia underwing, woody underwing, and Clinton's underwing. The well-camouflaged caterpillars of these moths, though seldom seen, provide vital protein-rich food for nestling birds.

Catocala adults are also difficult to see as they rest on trees during the day, covered by forewings that match the browns and grays of bark colors, often with a fringe on the edge of their wings to better disguise their outlines. But if touched or startled, the moths flash colorful hind wings as they fly off. It is believed this sudden flash of color scares off would-be predators.

One of the best ways to see and attract underwing adults is to bait them at night. Prepare a bait by mixing rotting fruit (bananas are often used), beer, and brown sugar and let it sit for twenty-four hours. Then smear the mix on tree trunks about eye height late in the day. Check the bait every hour at night, approaching as quietly as possible. The moths are easily startled by sound but are intent on feeding, so if you approach very cautiously you may be able to view and photograph these beautiful moths. A red cover on flashlights or headlights is less likely to frighten them off. You'll also discover other moths, beetles, and various night insects at your bait stations. For better inspection, you can net the moths and place them in clear plastic or glass containers. Be careful not to touch their fragile wings. Release them as soon as possible so they can get on with their lives and add to the insect diversity of the night landscape. Insects, particularly moths and butterflies, are declining dramatically, and experts warn that up to 40 percent worldwide are threatened with extinction.

Cockspur Hawthorn (*Crataegus crus-galli*)

Deciduous Tree
Size: Height 20–30 feet, width 20–35 feet
Color and Season: White flowers in late spring; bright orange to red foliage in autumn
Light Requirement: Sun, partial shade
Soil: Moist to dry; tolerable of clay, alkalinity, and occasional drought
Native Habitat: Savannas, open woodlands
Native to: AL AR CT DE FL GA IA IL IN KY KS LA MA MD ME MI MO NC NJ NY OH OK PA RI SC TN TX VA VT WI WV
USDA Hardiness Zones: 3–7

With a profusion of spring flowers, glossy green leaves in summer, spectacular orange-red fall foliage, and abundant red fruits that persist into fall and winter, this hawthorn offers beauty throughout the year. I grew it at my former home, but I don't grow it now because the red cedars at my current home would likely infect it with rust. Birds flock to the cockspur hawthorn for insects in spring and for red berries into late winter. With a short trunk and horizontal branching sweeping the ground, it is truly a gem of a specimen tree. Because of its fierce thorns, which are up to four inches long, you may want to avoid planting this where children play. A thornless variety is available.

Washington Hawthorn (*Crataegus phaenopyrum*)

Deciduous Tree
Size: Height 25–30 feet, width 20–25 feet
Color and Season: White flowers in late spring; orange to red foliage in autumn

Light Requirement: Sun, partial sun
Soil: Medium moisture; adaptable to many soils including those containing loam, some clay, sand, or rocky material
Native Habitat: Savannas, open woodlands
Native to: AL GA KY IL IN MD MO NC NJ NY OH PA SC TN VA WV; native but rare in FL LA MS; extirpated in AR
USDA Hardiness Zones: 3–8

This handsome tree is an all-season beauty with lovely spring blossoms, glossy green summer leaves turning brilliant red in fall, and red fruit that persists well into March. No wonder it is a favorite landscape tree. It is a hardy, durable tree and is more resistant to cedar apple rust than other hawthorns. Use it as a specimen tree or with several others in a hedge for privacy or partitioning. It was first discovered (by Europeans) near Washington, DC, hence *Washington* in its common name.

SUNFLOWER (*Helianthus* and *Heliopsis* species)

Aster Family, Asteraceae

Chapter 5 gives a full introduction to sunflowers and their outstanding value to birds and other wildlife. It features two of the most common sunflower species in prairies. This chapter features three additional sunflower species more common in savannas: the savanna sunflower (*Helianthus strumosus*), hairy sunflower (*H. hirsutus*), and woodland sunflower (*H. divaricatus*). It also includes a close relative, the false or oxeye sunflower (*Heliopsis helianthoides*), which resembles true sunflowers but produces seed in both the ray florets and disk florets of its flower heads, whereas true sunflowers produce seed only in their disk florets. All four have bright beautiful flowers that brighten savanna landscapes. Companion plants for these sunflowers include species more tolerant of the semishade of savannas. Their golden-yellow colors will look all the more brilliant when planted with pink roses, such as early or smooth rose (*Rosa blanda*), or blue asters, like sky blue aster (*Symphotrichum oolentangiense*), both of which are highlighted in this chapter.

An American Goldfinch perches on a sunflower.
PHOTO BY RYAN BRADY

The three *Helianthus* species are among a complex of several species that hybridize naturally, exhibiting considerable variation and overlap of characteristics, which make them difficult to distinguish. See a plant identification guide for details.

Woodland Sunflower (*Helianthus divaricatus*)

Perennial Wildflower
Size: 2–6 feet
Color and Season: Yellow flowers in summer to early fall
Light Requirement: Partial shade
Soil: Dry, medium dry; clay, loam, sand
Native Habitat: Savannas and associated prairies, open woods
Native to: AL AR CT FL IA IL IN GA KY LA MA MD ME MI MO MS NC NH NJ NY OH OK PA SC TN VA VT WI WV; native but rare in DE RI
USDA Hardiness Zones: 3–8

This species brightens savannas and prairies throughout much of the Midwest and is common in eastern states as well. It is easily distinguished from many other similar sunflowers by its stalkless or nearly stalkless leaves. The woodland sunflower can spread aggressively by rhizomes as do many sunflower species and is best planted in wilder areas where it's likely to outcompete invasive species, while adding golden color to the landscape.

Hairy Sunflower (*Helianthus hirsutus*)

Perennial Wildflower
Size: 2.5–5 feet
Color and Season: Yellow flowers in late summer to early fall
Light Requirement: Partial sun to full sun
Soil: Moist to dry; soil containing clay-loam, loam, rocky material, or sand
Native Habitat: Savannas and associated prairies
Native to: AL AR FL GA IA IL IN KS KY LA MI MO MS NC NE OH OK SC TN TX WI WV; native but rare in MA MI PA VA
USDA Hardiness Zones: 3–8

This species has also been dubbed bristly sunflower, rough sunflower, oblong sunflower, and stiff-haired sunflower. Perhaps these names have arisen from minor differences within the species, which has been divided into three varieties based on small differences in leaf size and hairiness—all of which testifies to the ever-evolving adaptations of this successful species. With a two-month blooming period, hairy sunflower offers nourishment for bees, butterflies, and other pollinators throughout much of the summer.

Savanna Sunflower (*Helianthus strumosus*)

Perennial Wildflower
Size: 3–6 feet
Color and Season: Yellow flowers in late summer
Light Requirement: Sun, partial shade
Soil: Moist to medium dry; clay, loam, sand

The savanna sunflower does well in dappled shade.
PHOTO BY MARIETTE NOWAK

Native Habitat: Savannas and associated prairies
Native to: AL AR CT IA IL IN FL GA KY LA MA MD ME MI MS NC ND NH NJ NY OH OK PA RI SC TN TX VA WI WV; native but rare in KS NE VT; extirpated in DE
USDA Hardiness Zones: 3–9

This lovely species is adapted to dappled shade and lightens up savannas in late summer with its bright yellow flowers. A common alternate name is pale-leaved sunflower, due to the lighter color of the undersides of the leaves. I grow this species under shrubs and trees, where it thrives in the muted light from an opening to the northeast. It can spread aggressively by rhizomes in ideal conditions. In contrast, it tends to wilt in dry soils in full sun and, alternatively, it fails to thrive in heavier shade. The savanna sunflower is one of the most variable of sunflowers and can be hard to distinguish from others. One key feature is its leaf stalks, which are one-quarter to more than an inch long. The other similar species have leaf stalks barely more than a half inch long (*H. hirsutis*) or nearly absent leaf stalks (*H. divaricatus*).

False or Oxeye Sunflower (*Heliopsis helianthoides*)

Perennial Wildflower
Size: 3–6 feet
Color and Season: Deep golden-yellow to orange-yellow flowers in summer
Light Requirement: Sun
Soil: Dry, medium, moist; clay, loam, sand
Native Habitat: Savannas, prairies, open woodlands
Native to: AL AR CO GA IA IL IN KS KY LA MD MI MN MO MS NC ND NE NJ NM NY OH PA SC SD TN VA WI WV; native but rare in DE OK
USDA Hardiness Zones: 3–9

Like many popular plants, this species has been given numerous names describing various aspects of the plant—early sunflower, false sunflower, smooth sunflower, and oxeye sunflower, among others. It tends to be a short-lived perennial and fades from restorations after a few years. But it self-seeds readily in bare ground and can adapt to nearly any soil from sand, loam, or clay, from moist to dry. The showy flowers bloom profusely throughout the summer. Even from a distance, you can often distinguish it from true sunflowers in the field by its deeper golden-yellow tones. In addition, both its ray florets and disk florets in its flower heads produce seed, whereas only the disk florets of true sunflowers produce seed. The seeds are square at the top, whereas true sunflower seeds are rounded. In many other ways, it resembles true sunflowers, as its specific epithet (the second word in its Latin name) denotes. *Helianthoides*

False sunflower has showy blossoms with deep golden tones.
PHOTO BY DAWN-MARIE STACCIA

means sunflower-like. Like true sunflowers, it attracts hummingbirds, many seed-eating birds, and a vast array of other fauna, including beneficial insects like syrphid flies and soldier beetles that help control aphids.

JUNIPER (*Juniperus* species)

Cypress Family (Cupressaceae)

> I would rate the tree forms of juniper as among the 10 essential plants to include in a bird-friendly landscape.
>
> —William Cullina, *Native Trees, Shrubs, and Vines*

Junipers are hardy, drought-tolerant plants, often underappreciated, despite their great value to an amazing variety of birds, insects, and other wildlife. Three of the most common species in the Midwest are featured here. Red cedar (*Juniperus virginiana*) is a tree and, although referred to as a cedar, it is a juniper species with beautiful reddish wood and would be better named red juniper. The most widespread shrub junipers are the common juniper (*J. communis*) and trailing (*J. horizontalis*).

Value to Birds and Other Wildlife

A bevy of birds feed on juniper's berrylike cones, which are fairly rich in both carbohydrates and fats. The "berries" are especially important as a winter food

Cedar Waxwings are named for their love of the berries of red cedar trees.
PHOTO BY ANNE MORETTI

for birds. The Cedar Waxwing is the most voracious feeder of the cones and has been named for the tree. Other birds especially fond of the berries include the American Robin, Eastern Bluebird, Eastern Phoebe, Evening Grosbeak, Northern Mockingbird, Pine Grosbeak, Purple Finch, and Yellow-rumped Warbler. The Yellow-bellied Sapsucker feeds on the sap.

Many other birds also feed on the cones, occasionally including the Alder Flycatcher, American Crow, Black-capped Chickadee, Blue Jay, Brown Thrasher, Chipping Sparrow, Cliff Swallow, Common Grackle, Eastern Kingbird, Fish Crow, Fox Sparrow, Gray Catbird, Hermit Thrush, Mourning Dove, Northern Bobwhite, Northern Cardinal, Northern Flicker, Ring-necked Pheasant, Ruffed Grouse, Swainson's Thrush, Tree Swallow, White-throated Sparrow, White-winged Crossbill, Wild Turkey, and the non-native *European Starling*. As they feed, most birds—like robins, jays, and mockingbirds—pick off the berries, toss their heads back, and swallow them whole, but Black-capped Chickadees pick out the seeds and let the pulp fall off.

Birds also use junipers for nesting. I found this is especially true in spring when deciduous trees are not yet leafed out. Species partial to red cedar as a nesting site include the American Robin, Blue Jay, Brown Thrasher, Cedar Waxwing, Chipping Sparrow, Common Grackle, Cooper's Hawk, Field Sparrow, House Finch, Northern Mockingbird, Northern Cardinal, Pine Warbler, Prairie Warbler, and Song Sparrow. The Chipping Sparrow and Northern Cardinal also nest in the common juniper shrub. I myself have enjoyed doves, cardinals, and jays nesting in my red cedars, and a cardinal nested in my common juniper shrub. The evergreen boughs of junipers also offer protective cover and roosting sites for the birds mentioned here, especially for wintering sparrows and owls.

A fascinating complex of insects, mostly as larvae, including forty-two moth and butterfly caterpillars, snack on various parts of junipers, providing food for birds. Quite a few of the insects are so closely associated with *Juniperus* species that they are named for their favorite food—the black-horned juniper borer, cedar borer, cedar tree borer, eastern juniper bark beetle, juniper aphid, juniper geometer moth, juniper hairstreak butterfly, juniper leafhopper, juniper midge, juniper plant bugs, juniper sawfly, juniper scale, juniper thrip, juniper tip midge, juniper webworm, northern cedar bark beetle, and the small cedar bark beetle.

Red cedar trees and juniper shrubs provide welcome cover for birds in winter.
PHOTO BY MARIETTE NOWAK

Birds and bugs aren't the only critters eating junipers. The black bear, gray fox, opossum, eastern chipmunk, and white-footed mouse all feed on the berry-like cones. White-tailed deer rely on red cedar for shelter from cold winter weather and sometimes browse on the leaves of the tree as an emergency food.

How to Grow

Junipers, including the red cedar, are easy to grow but need to be well watered the first season. Once established, they are extremely hardy and will thrive in clay and compacted soils as well as rocky and droughty soils. Avoid heavy shade, however, or soils that stay wet. Since junipers, especially red cedar, are tolerant of salt and air pollution, they are excellent species to plant on roadsides, unlike many other woody species. In winter, they tend to develop a burgundy color, but this is no cause for concern. They will green up again in spring.

Junipers can be aggressive in dry soils, and it is best to avoid planting them near sunny, dry prairies. In addition, they are very flammable, making it risky to conduct prescribed burns so vital to maintaining prairies. Junipers have male and female flowers on separate plants, so both need to be planted for fruit production. Cross-pollination occurs when the male flowers release their pollen into the wind and receptive female flowers receive it. Junipers are hosts for cedar apple rust and should not be planted within at least one hundred yards of apple, hawthorn, and crabapple trees.

Propagation

Seeds and cuttings.

Companion Plants

Junipers grow well with dry-tolerant shrubs like American hazelnut (*Corylus americana*) and fragrant sumac (*Rhus aromatica*). Their dark green foliage is also a beautiful backdrop to native grasses like big bluestem (*Andropogon gerardii*) and prairie dropseed (*Sporobolus heterolepus*) as well as native wildflowers such as gray goldenrod (*Solidago nemoralis*) and wild bergamot (*Monarda fistulosa*).

A Closer Look

In winter, look for female red cedar trees loaded with cone-like berries and you'll have a good chance of seeing a flock of Cedar Waxwings voraciously gobbling down the fruits. Waxwings are handsome crested birds, groomed in subtle brownish-gray tones, and named for the red, waxlike tips on their wing feathers. In northerly locations, you're more likely to find a flock of the similar-looking and closely related Bohemian Waxwings. Both species are nomadic in winter, traveling in large flocks in search of berries, but their movements are unpredictable, so you may have to keep an eye out throughout the season. Sometimes cedar fruits become overripe and start to ferment, producing alcohol. Waxwings gorging on such berries often become intoxicated and, while humorous to watch, can fall ill or even die.

When birds eat cedar fruits, the seeds are still intact in their droppings and thus are spread throughout wild areas. Take note of where red cedars are growing in fields. Often you'll find them along fence lines where perching birds may have "planted" the seeds.

Common Juniper (*Juniperus communis*)

Coniferous evergreen shrub
Size: Height 3–6 feet (sometimes up to 9 feet), width 12–20 feet
Color and Season: Inconspicuous flowers; fruit, a blue berrylike cone covered with white powder.
Light Requirement: Sun, partial sun
Soil: Moist to dry; well drained; acidic to alkaline
Native Habitat: Sandy soils on bluffs and dunes
Native to: AZ CA CT CO IA ID MA ME MI MN MT ND NE NH NM NV NY OR RI SD UT WA WI WY; native but rare in AL GA IL IN KY NC NJ OH PA SC VA VT WV; extirpated in DE MD
USDA Hardiness Zones: 3–6

Common juniper is a circumpolar species found throughout the temperate Northern Hemisphere and has the largest geographical range of any woody plant. It forms large sprawling mats of thick foliage, which is welcome cover for many birds and other wildlife in open areas. It can be distinguished from the other shrubby junipers by its needlelike leaves in whorls of three, unlike the scalelike leaves of trailing juniper. Both the foliage and the cones have a piney scent. Occasionally, under the right conditions, common juniper can develop into a larger shrub or even a small tree. There are two subspecies of common juniper that show some variations from one another.

Trailing Juniper (*Juniperus horizontalis*)

Coniferous evergreen shrub
Size: Height 2–12 inches, width 3–6 feet
Color and Season: Inconspicuous flowers; blue berry-like cone in autumn
Light Requirement: Sun
Soil: Moist to clay; well-drained acidic to alkaline
Native Habitat: Beaches and sandy prairies
Native to: MA ME MI MT ND SD WI WY; native but rare in AL IA IL IN MN NE NH NY OH VT
USDA Hardiness Zones: 3–9

Also called creeping juniper, this low-growing species can spread up to six feet, forming an attractive dense ground cover, making it ideal for foundation plantings, rock gardens, and edging. Unlike the common juniper, which has needlelike leaves, trailing juniper has scalelike leaves. It is native mostly in northern states but rare in many of them. During rainy springs, blight might cause some branches to brown, and they should be lopped off and discarded.

Red Cedar (*Juniperus virginiana*)

Coniferous Evergreen Tree
Size: Height 15–35 feet (sometimes up to 60 feet), width 3–12 feet
Color and Season: Cone-like flowers ripen into blue-colored seed cones in late summer
Light Requirement: Sun, partial sun
Soil: Moist to dry; nutrient-poor soil containing gravel, clay, sand, or rocky material
Native Habitat: Sandy savannas, rocky cliffs, dunes
Native to: AL AR CO CT DE FL GA IA IL IN KS KY LA MA MD ME MI MN MO MS NC ND NE NH NJ NY OH OK OR PA RI SC SD TN TX VA VT WI WV WY
USDA Hardiness Zones: 2–9

Red cedar is a widely distributed conifer, native to forty states. Because of its resistance to insects, the tree can live up to 300 years or more. One was aged at 795 years in Missouri! The tree's dense foliage not only offers cover and nesting sites for birds but also is very useful in landscaping as a screen or windbreak. The fragrant wood of red cedar repels insects and was commonly used to make cedar chests in the past; some are still made today, especially by the Amish. Its aromatic fruit is used as an ingredient in gin.

OAK (*Quercus* species)

Beech Family, Fagaceae

Oaks are the classic trees of savannas east of the Mississippi. Consequently, savannas are often referred to as oak savannas or oak openings. Bur oak (*Quercus macrocarpa*) is the iconic savanna species, but two other oak species often

An open, grown bur oak with branches arching to the ground.
PHOTO BY ANNE MORETTI

occur in savannas: black oak (*Q. velutina*) and Hill's oak (*Q. ellipsoidalis*). All three species will be discussed here, as one or the other might be best suited to your property's particular site conditions. The two main groups within the *Quercus* genus are white oaks and red oaks. Bur oak is in the white group and has round-edged leaves; black and Hill's oaks are in the red group and have leaves with pointed edges. In addition to the three major savanna oaks, several others occasionally occur in savannas: white oak (*Q. alba*), post oak (*Q. stellata*), blackjack oak (*Q. marilandica*), chinquapin oak (*Q. mulhenbergii*), pin oak (*Q. palustris*), red oak (*Q. rubra* or *borealis*), shingle oak (*Q. imbricaria*), and swamp white oak (*Q. bicolor*).

Value for Birds and Other Wildlife

> If you think of a plant as a bird feeder, which is exactly what it is, then in most regions, the oak makes the most food. In fact, oaks are ranked either number one or two in their support of the food web in 84 percent of all U.S. counties in which they occur.
>
> —Douglas W. Tallamy, *Nature's Best Hope*

Oaks host more than five hundred species of moth and butterfly caterpillars, providing a varied and nutritious smorgasbord for birds throughout the year.

Migrating warblers flock to their branches in spring to plump up after their strenuous cross-country trips. They feed on the inchworms and other caterpillars busy nibbling on oak leaves as the insects emerge from eggs or hibernation. Research has recently confirmed the extraordinary value of oaks for warblers and other migrants. Christy Beal, a student of entomologist Doug Tallamy, compared the number of minutes warblers foraged among fifteen different genera of trees during spring migration in New Jersey. She discovered that the birds foraged three times longer in oaks than in pines (the next closest competitor), six times longer in oaks than in birches, and very little time in twelve other tree genera.

Throughout the growing season, birds are supplied with many other treats, including leafhoppers, ants, gall insects, plant bugs, and beetles to fatten their nestlings. The great leafy boughs and varied bark textures of oaks offer shelter for birds and bugs alike. Cavities in older trees are used by screech-owls, woodpeckers, and other cavity-nesting birds, as well as squirrels.

In autumn, the large acorns of oaks are a superfood for wildlife, rich in protein, carbohydrates, and fat, as well as essential minerals. Dozens of bird species in the eastern United States eat acorns, including Blue Jays, Mallards, Wood Ducks, Tufted Titmice, White-breasted Nuthatches, Woodpeckers, Thrushes, Towhees, Grackles, Thrashers, Northern Bobwhites, and Ruffed Grouse. Acorns are an important diet staple for many mammals, too, including squirrels, raccoons, opossums, bears, mice, chipmunks, and deer. Even some insects specialize in acorns, such as weevils, ants, and an acorn moth.

Later in fall, the leaf litter beneath oaks provides cover for ants, gall insects, spiders, and a myriad of other insects, as well as salamanders, frogs, mice, and other small critters. All this life in the oak leaf litter offers a bonanza of healthy treats for ground-feeding birds, like flickers, towhees, sparrows, and juncos. The leaf litter also reduces the germination of invasive species, thus ensuring the domination of the native plants useful to birds.

And since oaks are some of our longest-lived tree species, these feasts for wildlife are a reliable staple for many decades, even centuries. On top of that, oaks are champions of carbon sequestration, storing carbon in their tissues throughout their long lives and helping mitigate the detrimental effects of climate change on birds (and us).

Because of their outstanding value to wildlife, oaks are considered one of the top keystone species. Keystones at the top of an arch lock the other stones in place; if removed the arch falls. Similarly, without oaks and other keystone plants, our ecosystems would likely fail.

How to Grow

Some oaks have deep taproots, making them difficult to plant by the time they are several years of age. For this reason, it is often best to plant them as acorns or seedlings. Although relatively slow growing, oaks are long-lived, and if properly planted and protected when young, they can reach 100–300 years of age, depending on the species. See the sidebar Planting an Oak for more information.

Oak wilt is one of the greatest threats to oaks in the Midwest and Northeast, affecting both the red and white oaks. It is more severe in red oaks and kills them more quickly. Oak wilt is caused by a fungus (*Ceratocystic fagacearum*), which is spread by beetles that feed on the sap in open wounds. To prevent this disease, avoid pruning or otherwise injuring the trees during the growing season from April to October. Also avoid transporting diseased oak wood. If a tree must be pruned during the growing season due to storm damage, seal the wound immediately with a tree wound sealant or latex paint. Outbreaks of spongy moths (formerly called gypsy moths) can also be devastating for oaks. To help oaks resist the damage from the caterpillars, water yard trees weekly during dry conditions. Healthy oaks can regrow leaves within a few weeks after such outbreaks.

Propagation

Seeds (acorns).

Companion Plants

Some of the herbaceous plants mentioned in this chapter commonly grow under savanna oaks, including calico aster (*Symphyotrichum lateriflorum*), elm-leaved goldenrod (*Solidago ulmifolia*), and savanna sunflower (*Helianthus strumosis*). A hummingbird favorite, wild bergamot (*Monarda fistulosa*), also grows well in the partial shade of savannas.

A Closer Look

Oaks are abundant in the Midwest where I live, especially on the morainal lands left by the Wisconsin glaciation, which occurred between 31,000 and 11,000 years ago. The first trees to populate our bare soils after the glaciers melted were spruces and firs, trees adapted to cold climates. But as the climate warmed, oaks gradually replaced the conifers. How did they get here? Squirrels have a range of only a few acres within which they bury acorns in fall to store for the winter. They forget some, which eventually grow into oaks, but the pace at which they spread oaks is quite limited. Blue Jays, however, bury acorns over a mile away from an oak tree, and they bury masses of acorns, as many as three to five thousand each fall. Like squirrels, they fail to retrieve some, up to a quarter of them each year. The "airlifting of oaks," as some researchers have described the Blue Jay's dispersal of acorns, easily accounts for the rapid spread of oaks after glaciation.

Planting an Oak

Planting an oak is in a sense like sowing the seeds of your own immortality by helping something to take root that will likely live for 200 years or more.

—William Cullina, *Native Trees, Shrubs, and Vines*

1. Choose an appropriate site. The site should be large enough for a mature oak(s), with well-drained soil and at least a half day of sun.
2. Choose an oak species native to your region and adapted to the soil, light, and other conditions of the site.
3. Consider planting a cluster of two or three oaks. These oaks will intertwine roots, giving them more stability and ensuring that they will be less likely to topple in severe weather. Plant about ten feet apart to mimic the way they'd grow together in woodlands. Savanna oak species. however, sometimes occur singly or at greater distances from one another.
4. Plant acorns or seedling oaks, rather than established trees, to produce healthier trees.

Planting an acorn

For best-quality acorns, collect in fall when they are brown. Plant acorns in the white oak group immediately, since they grow roots in fall. Acorns in the red oak group can be either planted in fall (although they will not germinate until spring) or placed with damp peat moss or sawdust in a refrigerator until planting in spring.

In either case, plant in a deep pot to accommodate the initial taproot. Use a mixture of local soil and potting mix. Cover with chicken wire to protect from critters. If planted in autumn, water about once a month so they don't dry out, and place the pot in a sheltered spot to avoid the worst of winter weather.

Planting a 1- to 2-year-old seedling

Seedlings, often in the form of bare root whips, are small trees a few feet tall. Plant in a hole a little wider than the spread of the roots, but no deeper than the crown (the place where the first roots appear). The whip will be killed if planted too deep. Water liberally.

Early care

Water once or twice a week as needed during the growing season.

Protect with chicken wire, gradually increasing its size as the trees grow. Eventually, to prevent deer browsing, a wire cage about five feet tall and wide enough for the spread of the oak branches will be needed in the early years of growth.

A Closer Look

Even young oak seedlings and saplings contribute to the food web. I've noted that their young leaves are often punctured with holes, suggesting that they are especially tender and palatable to insects, who in turn provide nutritious offerings for birds and other wildlife up the food chain. Young oaks can usually survive some insect damage but need protection from deer and other munching mammals.

These guidelines are based on those in Tallamy's *The Nature of Oaks*, on other experts' advice, and on personal experience.

Hill's or Northern Pin Oak (*Quercus ellipsoidalis*)

Deciduous Tree
Size: Height 50–75 feet, width 35–60 feet
Color and Season: Yellow-green flowers in catkins; scarlet foliage in autumn
Light Requirement: Sun, least shade-tolerant oak
Soil: Tolerates drought; requires sandy, well-drained soil
Native Habitat: Savanna, morainal hills, open woodlands
Native to: IA IL IN MI MN ND OH WI; rare in MO
USDA Hardiness Zones: 4–7

Hill's or northern pin oak is a medium-sized oak, smaller in stature and width than the bur and white oaks, so it is better suited for urban or suburban properties. We planted this species on our quarter-acre property in a suburb of Milwaukee where we previously lived. It became the centerpiece of a small savanna in the corner of our yard. Like so many homeowners, we first planted it smack in the middle of lawn grasses. But over time, as we learned more about native landscaping, we planted flowers and grasses native to savannas and tolerant of the partial shade below the tree. Since this species has a taproot, like many oaks, we were able to do this without disturbing its root. And voila! We had a miniature savanna right in our own yard. It should be noted that experts widely disagree on the soil tolerances of the northern pin oak, but ours did very well in the clay soil on our property. Also important to bear in mind: this species is in the red oak group and is more susceptible to oak wilt. Care should be taken to avoid pruning during the growing season. This oak was a beautiful addition to our yard and, more importantly, offered the most wildlife value of any species we had introduced onto our property. We attracted more than eighty species of birds to our small yard, likely due, in no small part, to the pin oak that we had planted.

Bur Oak (*Quercus macrocarpa*)

Deciduous Tree
Size: Height 40–80 feet (sometimes to 120 feet), width 40–60 feet
Color and Season: Yellowish-green flowers in catkins in spring; golden-yellow-brown foliage in autumn
Light Requirement: Sun to partial shade
Soil: Wet to dry, except flooded site; tolerant of wide variety of soil types: sands, clays, loams
Native Habitat: Savannas, moist bottomland woodlands, upland woodlands
Native to: AR IA IN IL KS KY MD ME MI MN ND NE NY OH OK PA RI SD TN TX VT WI WV WY; native but rare in AL CT LA MA MD MS MT NH NM VA VT
USDA Hardiness Zones: 3–8

Bur oak is one of my favorite trees, with branches arching nearly to the ground in open savannas and a crown as wide as high. It has rugged, noble presence—a truly mighty oak. It's ridged, corky bark protects it from fires that were once common in the original oak savannas and are today replicated with prescribed burns. The acorns are the largest of the native oaks and have a fringe around the rim of the cup, for which the tree has been named *bur* oak.

I'm fortunate to have several bur oaks in my yard. They provide welcome shade for me and leafy havens for birds in summer, plus impressive architectural

This Eastern Kingbird has built a nest in the boughs of a bur oak.
PHOTO BY JACK BARTHOLMAI.

interest in the winter. Bur oaks are hardy trees, resistant to salt, pollution, and drought and adaptable to a variety of soils, including highly alkaline soils. Their deep taproots make them difficult to transplant so the trees are best planted as acorns or seedlings one or two years old (see the sidebar Planting an Oak). Native plant nurseries also offer balled and burlapped bur oaks, which are relatively easy to plant. Although fairly slow growing, bur oaks are extraordinary trees that will add to your enjoyment through the years and provide a legacy for the future. They should be staples, from urban to rural properties, wherever there is sufficient room to accommodate their mature size. Regardless of where they are planted, oaks increase property values.

Black Oak (*Quercus velutina*)

Deciduous Tree
Size: Height 75–100 feet, width 60–75 feet
Color and Season: Yellow-green flowers in catkins in spring; yellow to red foliage in autumn
Light Requirement: Sun to partial shade
Soil: Dry to medium; dry sandy soils to rich silty soils, even clay soils, intolerant of alkalinity
Native Habitat: Savannas, barrens, open woodlands
Native to: AL AR CT DE FL GA IA IL IN KS KY LA MD ME MI MN MO MS NC NE NH NJ NY OH OK PA RI SC TN TX VA VT WI WV
USDA Hardiness Zones: 3–9

This is a major oak species in sandy savannas. In the Kettle Moraine area of southeastern Wisconsin where I live, the glaciers deposited a great deal of gravel and sand close to the soil surface, creating sites for dry, sandy savannas where this hardy species thrives.

Black oak is in the red oak group and more susceptible to oak wilt. Care should be taken to avoid pruning during the growing season.

Black oaks host a fascinating array of ant species, likely more than any other oak species. Some of the ants are exceedingly rare, a few even unnamed. The ants can be found in all parts of the trees—in the branches, bark, and leaves of

A Northern Flicker brings food to its young nestled in a cavity of an oak tree.
PHOTO BY JACK BARTHOLMAI

living trees and in snags (dead standing trees) and logs. They often nest in the acorns. With this rich bounty of ants, it's no wonder that Northern Flickers are common birds of savannas. These woodpeckers eat more ants than any other species of birds, lapping up the insects with their sticky tongues. Northern flickers are known to consume more than five thousand ants in one sitting alone!

ROSE (*Rosa* species)

Rose Family, Rosaceae

Roses are one of the most common shrubs found in remnant savannas, so it is especially fitting to include them in savanna gardens and restorations today. Unlike most cultivated roses, our native wild roses are easy to grow, drought resistant, and winter hardy. All have lovely fragrant flowers with a single circle of petals, providing pollinators easy access to pollen (roses don't produce nectar).

The fragrant flowers of native roses have a single circle of petals, providing pollinators easy access to pollen.
PHOTO BY MARIETTE NOWAK

The many petals of hybrid and cultivated roses often prevent pollinators from accessing their pollen. Some have even lost all their pollen and sometimes their scent, too, as a result of their breeding. For native gardeners, roses as well as other plants should be more than "pretty faces." They should play a role in the ecosystem and provide a service to wildlife. Native roses provide for dozens of wildlife species, from birds and bees to moths and mammals. Their beautiful simple "faces" are more than just ornaments in our gardens.

Three commonly grown savanna roses are highlighted here: the early or smooth rose (*Rosa blanda*), pasture rose (*R. carolina*), and climbing wild rose (*R. setigera*). Two other roses deserve mention but are not usually found in savannas. Dwarf prairie rose (*R. arkansana*) is native to much of the Midwest and Great Plains but prefers the full sun of prairies. Swamp rose (*R. palustris*) is native in eastern states and parts of the Midwest and thrives in swamps and marshes.

Value to Birds and Other Wildlife

Roses offer a feast for birds year-round. During the growing season, the many insects attracted to roses, including 135 species of caterpillars, provide a banquet for birds. Later, in winter, the berrylike rose hips offer important nourishment during lean times. Roses also provide cover and nesting sites for many birds.

Rose hips are a favorite food of the Cedar Waxwing, Northern Mockingbird, and Swainson's Thrush. Other birds that feed on the hips are the American Goldfinch, American Tree Sparrow, American Robin, Brown Thrasher, Dark-eyed Junco, Eastern Bluebird, Evening Grosbeak, Fox Sparrow, Gray Catbird, Greater Prairie Chicken, Northern Bobwhite, Northern Cardinal, Philadelphia Vireo, Song Sparrow, Wild Turkey, and Wood Thrush. The Greater Prairie Chicken, Northern Bobwhite, Ring-necked Pheasant, and Ruffed Grouse munch on both the hips and buds.

With their dense branching and protective thorns, roses are also popular nesting sites for birds. Brown Thrashers, Gray Catbirds, Northern Cardinals, Northern Mockingbird, and Song Sparrows are especially likely to build their nests in roses. To my delight, cardinals have chosen to nest in my climbing wild rose several times, their nests quite well hidden in the tangle of the vines. Rose bushes also provide cover for a variety of birds.

Among the caterpillars hosted by roses are those of geometer moths, midget moths, owlet moths, prominent moths, sphinx moths, trumpet leaf miner moths, and tortrix moths. Another, the stinging rose caterpillar, is a colorful species that can inflict a painful sting. The invasive spongy moth (formerly called gypsy moth) caterpillar also commonly devours rose leaves. These bristly caterpillars are eaten by larger birds like cuckoos, Blue Jays, orioles, and towhees—but obviously not often during outbreak years.

More than twenty-five insects are named after their favored food plant: bristly rose slug, curled rose sawfly, fuller rose beetle, imported rose sawfly, lesser rose aphid, rose aphid, rose chafer, rose curculio, rose-grass aphid, rose hip maggot, rose midge, rose plant bug, rose scale, rose shoot sawfly, rose slug, and yellow rose aphid. The others include seven rose gall wasps, the most striking of which is the pink spiny rose gall, quite showy when fresh. In addition, many bees visit the flowers, vibrating their flight muscles to dislodge the pollen. Other insects from grasshoppers to weevils munch on various parts of rose plants. All are potential bird food. And despite all these insects, there is only occasional damage of any significance reported on native wild roses, minimized surely by the bevy of birds snacking on the insects. Cultivated roses, however, unless grown in ideal climates, like those along the West Coast of our country or in England, are often extremely susceptible to insect damage as well as diseases and winter kill, although some cultivars have been bred to address some of these problems.

Besides birds and insects, small mammals also are fond of rose hips. And surprisingly, cottontail rabbits and white-tailed deer browse on the stems, despite their prickles and stout thorns.

How to Grow

The roses featured here grow well in the light shade of savanna-like settings but also grow in sunny prairies. They prefer rich soils but will tolerate poorer soils. Early rose and pasture rose can be grown in hedges, mounds, and woodland borders. These shrub roses tend to sucker but can be controlled by heading back, or using a spade around the periphery of the shrub to sever any suckers each spring. Climbing wild rose is an excellent choice for a trellis or arbor and does not sucker like the shrubs.

Propagation

Cuttings and seeds.

Companion Plants

Whether shrub or vine, the beauty of our native roses is best framed with the green of grasses or sedges at their bases, setting off their beauty without the need for any competing colorful wildflowers. They are also stunning against evergreen backgrounds. My climbing wild rose is in a dry, sunny spot with buffalo grass (*Buchloe dactyloides*) at its base, a lovely low grass that needs no mowing. Prairie dropseed grass (*Sporobolus heterolepus*) and common oak sedge (*Carex pensylvanica*) are other attractive options that do well in sun or light shade on moist to dry soils.

A Closer Look

Summer is the time to enjoy the flowers and pollinators. Note the numerous yellow stamens loaded with pollen in the center of each rose flower—a welcome signpost for pollinators. It's no wonder that bees often swarm around native rosebushes. I'm not a bee expert, but I like to listen to their buzzing and count how many different kinds of bees are visiting. The more diversity the better during these times when biodiversity is diminishing. Also, look for the spiny rose gall, a lovely reddish gall when fresh and circled with spines that seem to mimic those of their host plant. It's most often found on the pasture rose. I've been lucky to see this gall once and hope to do so again.

In winter, watch for birds like mockingbirds and Cedar Waxwings devouring rose hips, one of their survival foods when little else is available. These birds specialize in feeding on the hips from late fall through winter. You may spot them feeding on roses on Christmas bird counts and winter walks or while skiing.

Pasture Rose (*Rosa carolina*)

Perennial Shrub
Size: Height 1–2 feet, width 3–4 feet
Color and Season: Pink to white flowers in summer; yellowish to orange-and-red foliage in autumn
Light Requirement: Sun to partial shade
Soil: Dry to moist; loam, sand
Native Habitat: Savannas, prairies
Native to: AL AR CT DE FL GA IA IL IN KS KY LA MA MD ME MI MO MS NC NE NH NJ NY OH OK PA RI SC TN TX VA VT WI WV
USDA Hardiness Zones: 4–9

This species is one of the shorter rose shrubs and has one of the widest native ranges, extending from the eastern states through the Great Plains. Although it is one of the most shade-tolerant native roses, it grows best with at least partial sun. Unlike the smooth rose, it is a bristly species. It blooms for about a month with showy two- to two-and-a-half-inch flowers. Like the other roses featured here, its flowers are darker pink at first, then fade to light pink over the next few days. The small size makes it an excellent choice for savanna gardens.

Early or Smooth Rose (*Rosa blanda*)

Perennial Shrub
Size: Height 2 to 6 feet, width 4–6 feet
Color and Season: Pink to white flowers in early summer; yellow foliage in autumn
Light Requirement: Sun to partial sun
Soil: Dry to medium; clay, loam, sand
Native Habitat: Savannas, open woodlands, prairies
Native to: AR CT IA IL IN KS ND NE NH NY MA ME MI MN SD VA VT WI; rare in MD MO OH PA WV
USDA Hardiness Zones: 3–7

The common names of this rose are fitting. It's one of the earliest native roses to bloom and, unlike other roses, it has mostly smooth stems with very few thorns. It can spread rapidly, forming large colonies in good conditions and making it ideal for thickets and borders. Suckering may need to be controlled if the spread is undesired, by heading back with a spade each spring.

Climbing Wild Rose (*Rosa setigera*)

Perennial Vine
Size: Height 6–15 feet, width 5–10 feet
Color and Season: Pink flowers in early to midsummer; yellow to red, then purple foliage in autumn
Light Requirement: Sun, partial sun
Soil: Dry, medium, moist; clay, loam, sand
Habitat: Savannas, open woodlands
Native to: AL AR IA IL IN KS KY LA MI MO NY OH OK TN TX WV; native but rare in FL GA MS NC NE PA SC VA
USDA Hardiness Zones: 5–8

The climbing wild rose displays a profusion of flowers.

PHOTO BY MARIETTE NOWAK

Although sometimes called prairie rose, this species prefers savannas and woodland edges with half-day sun. It can be grown as a mound or trained to grow as a vine. It is an excellent choice for a trellis where the multitudes of flowers are on full display. The blossoms are darker pink when they open and gradually become light pink as they age, with new ones opening as others fade, thus creating a cascade of many colors that lasts for about a month. It's my favorite rose, not only beautiful to see but harmonious to hear as well when in bloom and abuzz with bees. It blooms on new wood, so cut back in early spring to promote new growth. Unlike many shrub roses, this species does not sucker but does best with renewal pruning (thin out older canes to base) every three years. Also unique, this rose is dioecious, unlike most other roses, with the male and female flowers occurring on different plants. Both sexes are needed to produce rose hips. Since the sex of young immature plants cannot be determined, install three to five plants fairly near each other to increase the chances of having both sexes.

LITTLE BLUESTEM GRASS (*Schizachyrium scoparium*)

Grass Family, Poaceae
Perennial Grass
Size: Height 2–4 feet, width 1 foot
Color and Season: Purplish-bronze flowers in summer, reddish-bronze foliage in fall
Light Requirement: Sun

Little bluestem has a reddish-bronze fall color and fluffy seeds with long white hairs.
PHOTO BY MARIETTE NOWAK

Soil: Dry to medium; loam, sand
Native Habitat: Savannas, prairies, woodland edges
Native to: All lower 48 states except CA NV OR; rare in ID WA
USDA Hardiness Zones: 3–10

Little bluestem is an iconic grass of the original savannas and prairies that once graced much of North America. It is my personal favorite short native grass, with beautiful foliage that is usually blue green in summer and radiant reddish bronze in fall, a lovely contrast to the abundant white-tufted seeds all along the stems. It should be noted that the foliage color can be highly variable. While usually blue green, it ranges from bright green to a glaucous bluish gray, and sometimes these variations can be found at the same site, as is true on my own property.

Value to Birds and Other Wildlife

The Eastern Towhee sometimes builds its nest at the base of a little bluestem clump on the soil and dense surface roots, with the foliage arching over to conceal the nest. The seeds are a principal source of food for songbirds, such as Field Sparrows, Tree Sparrows, and juncos, especially in winter. The foliage is used for nesting material and welcome cover amid summer storms and winter snows. Many of the insects described here are essential protein sources for birds.

The most nutritious food sources for birds are the caterpillars of the many skippers and grasshoppers that feed on the foliage of the grass. The skippers include the dusted skipper, cobweb skipper, Ottoe skipper, Indian skipper, swarthy skipper, and crossline skipper. Other insects that feed on little bluestem include the prairie walking stick, a leaf-mining beetle, leafhoppers, thrips, plant hoppers, and spittlebugs. Bison, cattle, and other hoofed mammals thrive on the grass.

How to Grow

This beautiful grass, with exceptional color in all seasons, is often used in both restorations and landscaping. It is a clump-forming grass that emerges slowly in spring. It is easy to grow, preferring dry to medium-sandy soils, but it tolerates a variety of soils, with the exception of heavy soils that retain moisture. Little bluestem is perfect for savanna and prairie gardens where short species may be preferred. It's also useful for erosion control on slopes and riverbanks. It readily self-sows, so gardeners of small sites may want to remove the seeds (but this will eliminate a seed source for birds).

Propagation

Seeds, division.

Companion Plants

The blue-green foliage is an attractive complement to sunflowers (*Helianthus* species), blazing stars (*Liatris* species), and asters (*Symphyotrichum* species).

A Closer Look

Watch for skippers rapidly darting about in gardens and prairies with little bluestem grass. If you're lucky, you may see them lay an egg on the foliage. Skipper caterpillars are some of the most common insects feeding on little bluestem. True butterflies and skippers are both considered butterflies but are in different superfamilies (Papilionoidea and Hesperioidea, respectively). As adults, skippers have larger bodies in proportion to their wings than true butterflies, have relatively smaller wings, and usually have a club at the end of their antennae that is hooked backward like a crochet hook, unlike the simple clubbed antennae of true butterflies. When basking, skippers usually hold their wings flat, although some hold their wings closed or partially open in a V or U form.

This crossline skipper is holding its wings partially open.

PHOTO BY MIKE REESE, wisconsinbutterflies.org

As adults, skippers are mostly drab brown and/or orange in color and are notoriously difficult to identify. If you, like me, are not up to the challenge, simply enjoy their fast, skipping flight for which they've been named. You may also want to search for the skipper caterpillars mentioned here. They have a variety of colors and can be easier to identify than the adults. The dusted skipper caterpillar is pink lavender on top with pale gray sides and is covered with long yellow-white hairs. Its head is reddish purple. The cobweb skipper caterpillar is light brown with a green dorsal stripe along its body and a black head. The Ottoe skipper caterpillar is greenish brown with a dark brown head. The mature Indian skipper caterpillar is reddish brown mottled with green highlights and light speckling. The swarthy skipper caterpillar is white with a brown head. The crossline skipper caterpillar is dark brown with a black head and whitish marbling.

You'll likely see birds searching for these caterpillars—rich protein morsels for them and their nestlings. And if you're especially fortunate, you might find a towhee's nest beneath the grasses. In late fall and winter, watch for the many birds that feast on the seeds and seek cover under the grasses in bad weather.

ELM-LEAVED GOLDENROD (*Solidago ulmifolia*)

Aster Family, Asteraceae
Perennial Wildflower
Size: 2–4 feet
Color and Season: Yellow flowers in late summer and fall
Light Requirement: Light shade, partial sun
Soil: Medium to medium dry; loam, sandy loam, glacial till, some rocky material
Native Habitat: Savannas, open woods
Native to: AL AR CT IA IL IN KS KY LA MA MD MI MN MO MS NE NH NJ OK PA RI TN TX, VA WI WV; native but rare in DE, FL, GA, NC, SC, VT; extirpated in ME
USDA Hardiness Zones: 3–8

Chapter 5 offers a general introduction to goldenrods and their great value to birds and other wildlife. It features two goldenrod species common in prairies that will also do well in sunnier areas of savannas. Another species, the elm-leaved goldenrod, does well in the moderate shade of savannas. Companion

The handsome elm-leaved goldenrod grows well under the shade of oaks.
PHOTO BY MARIETTE NOWAK

plants include those that also do well in moderate shade. Pink roses, such as early or smooth rose (*Rosa blanda*), and blue asters, like sky blue aster (*Symphotrichum oolentangiense*), both of which are highlighted in this chapter, will contrast beautifully with the yellow flowering panicles of this attractive goldenrod. Like all goldenrods, it attracts droves of insects that offer high-protein feasts, especially many fat, juicy caterpillars, for savanna birds and their nestlings.

Elm-leaved goldenrod has a graceful, vase-shaped panicle of flowers that reminds me of the arching branches of the American elm trees that lined the streets when I was a child. It also has wide lower leaves that resemble those of the elm after which it is named. A handsome species, it grows very well under the shade of oaks on the morainal slopes of my small woodland. Like all goldenrods, it is beloved by a wide array of bees and butterflies seeking its nectar and pollen.

AMERICAN ASTER (*Symphyotrichum species*)

Aster Family, Asteraceae

Chapter 5 offers a general introduction to American asters and their outstanding value for birds and wildlife. It features three aster species common in prairies. Here I describe two additional aster species, the calico aster and the sky blue aster, both of which do well in the moderate shade of savannas. Excellent companion plants for these asters include the other herbaceous species highlighted in this chapter, including little bluestem grass (*Schizachyrium scoparium*), elm-leaved goldenrod (*Solidago ulmifolia*), and savanna sunflowers (*Helianthus* and *Heliopsis* species). Like all asters, both attract multitudes of insects that offer high-protein feasts, especially many fat, juicy caterpillars for savanna birds and their nestlings.

A Closer Look

Closely examine your asters and you may discover a woolly bear caterpillar (*Pyrrharctia isabella*), one of the best known of the many moth caterpillars that feed on asters. It has a reddish-brown band in the middle of its body with black bands front and back. According to a popular myth, the severity of the coming winter can be predicted by the relative length of the brown bands. In actuality, the color variations are primarily due to the age of the caterpillars, with older

The calico aster has multitudes of small flowers on wide-spreading branches.
PHOTO BY MARIETTE NOWAK

ones having wider brown bands. The caterpillars wander about in the fall to find a suitable place in the leaf litter to hibernate for winter. Beige hairy cocoons are made in spring and again in summer by two generations of woolly bears. The Isabella tiger moth, a two-inch yellow-brown moth, emerges from the cocoons.

The caterpillars, sometimes called woolly worms, are common throughout the United States and into Canada and Mexico. Many towns celebrate their appearance, and woolly worm races are a highlight of fall festivals. The fastest "worm" predicts the winter weather.

Calico Aster (*Symphyotrichum lateriflorum*)

Perennial Wildflower
Size: 1–3 feet
Color and Season: White flowers in late summer / early fall
Light Requirement: Part shade to shade
Soil: Dry, medium; clay, loam, sand
Native Habitat: Savannas, woodlands
Native to: AL AR CT DE IA IL IN FL GA KY LA MA MD ME MI MN MO MS NC NY NJ NY OH OK PA RI SC TN TX VA VT WI WV; native but rare in KS NE SD
USDA Hardiness Zones: 4–9

Often called side-flowering aster, this species has clusters of small composite flowers primarily on the upper side of its wide-spreading branches. The *calico* name comes from the varied colors of the flowers' central disks, which range from fresh yellow to purple as they fade and are pollinated. A single plant often has disks in a variety of color stages, giving it a calico look. This attractive aster has a long bloom time of one to two months. A beautiful patch of these asters grows at the edge of a kettle pond in my yard.

Sky Blue Aster (*Symphyotrichum oolentangiense*)

Perennial Wildflower
Size: 2–3 feet
Color and Season: Blue flowers in late summer / early fall

The sky blue aster blooms in profusion for a month in late summer and early fall.
PHOTO BY MARIETTE NOWAK

Light Requirement: Sun to partial shade
Soil: Dry to moderate moisture; loam, sand
Native Habitat: Savannas, prairies
Native to: AR FL IA IL IN KS LA MI MO MN MS NE OH OK WI TX; native but rare in AL GA KY NY SD TN
USDA Hardiness Zones: 3–8

Formerly called azure aster, the sky blue aster has showy blue-violet flowers that bloom in profusion for a month late in the season. Unlike the smooth leaves of the similar smooth aster, the leaves of sky blue aster feel as rough as sandpaper, a handy way to separate the two in the field. Sky blue aster is easy to grow and drought tolerant and does well in any kind of soil that is well drained. This species is a great addition to any garden or restoration, especially one that tends to be on the dry side.

Indigo Bunting.

PHOTO BY JACK BARTHOLMAI

7 Top 10 Plants for Indigo Buntings and Other Shrubland Birds

Brilliant Indigo Buntings, Eastern Towhees, and Gray Catbirds are among the many handsome and fascinating birds that feed, sing, court, and nest in shrubs, spending nearly their entire lives within the welcoming shelter of leafy shrubbery. Native shrubs provide them with every necessity—food, protection from predators and storms, and sturdy nesting and perching sites.

Unfortunately, too much of our shrubbery today is chock full of non-native invasive species like buckthorn, honeysuckle, multiflora rose, and barberry. These invasive shrubs often become monocultures, wiping out native shrubs that are the best species for providing birds with the right foods at the right times in their life cycles. Studies also have confirmed that birds nesting in invasive buckthorn and honeysuckle shrubs are more apt to be victims of predators.

But one of the worst problems with all non-native shrubs, whether invasive or cultivated, is that they don't support insects that are so essential for birds. As mentioned earlier, our insects have not evolved with these foreign shrubs and don't have the necessary enzymes to feed on them. The result is a food desert for insects and for the birds that rely on them for some or most of their food. The majority of land birds, for example, require insects, especially fat, juicy caterpillars, to feed their nestlings, which starve or die without the food they need, as shown in research described in chapter 1. Other species of birds are insectivores, feeding only on insects throughout their lives. Because of these problems, non-native shrubs are considered "sinks" for birds and cause significant declines in bird populations. Unknowingly, we have been starving our birds by planting non-native cultivated plants!

The good news is that there are many excellent native shrubs that provide birds with a healthy variety of nourishing berries, seeds, nuts, and protein-packed insects. This chapter features some of the best species for birds—shrubs that can replace cultivated species that offer little for birds. For example, consider replacing forsythia, which hosts only a single caterpillar species, with a willow shrub in a genus that supports 456 species of caterpillars. Replace weigela, on which only 2 caterpillar species are found, with ninebark, which supports 41 species. Or replace cotoneaster, on which only 3 caterpillar species can survive, with dogwood, which supports 118 species. The other native shrubs described here will suggest additional species that may be suitable for your property and excellent replacements for various non-native cultivated shrubs as well as for invasive shrubs like buckthorn and honeysuckle.

Most homes have a shrub or two around their foundations that are apt to be perfect sites for native shrubs that do well in drier soils, such as St. John's wort or hazelnut. Many yards also have a swale or drainageway along property boundaries where shrubs requiring more moist soils, such as red-twigged dogwood and nannyberry viburnum, will thrive. Border shrubs also offer privacy and a backdrop for herbaceous vegetation. Many such swales run through an entire neighborhood, and ideally all properties could harbor a corridor of native shrubs, allowing the natural movement of birds and other wildlife through communities. Consider planting three of each species, if possible, for better pollination. And install in a zigzag pattern for a more natural look.

Also consider replacing monoculture hedges with a "bird hedge," using a variety of native shrubs for beautiful all-season interest. Picture clusters of fragrant white chokecherry blossoms in spring followed in summer by a riot of blooms blanketing dogwoods and viburnums, each in their time, along with the large pancake-like clusters of American elderberry blossoms. The bright yellow bee-loved blossoms of St. John's wort usually come onto the stage as well. Late summer brings bright berries in red, blue, or white on many of these shrubs, along with the reddish-brown capsules of St. John's wort seeds and the hidden nuts of hazelnuts. By fall, beautiful leaf colors of pumpkin, cherry red, lemon yellow, and burgundy offer a stunning display. Winter brings a variety of interesting branching structure and color, especially when highlighted by snow.

Birds to Watch for in Your Shrubland Garden

The following are 32 species of birds that breed in shrubby habitats in Eastern North America:

Northern Bobwhite

Indigo Bunting

Northern Cardinal

Gray Catbird

Flycatchers: Alder and Willow

American Goldfinch

Blue Grosbeak

Sparrows: Clay-colored, Field, Lark, Song, Lincoln's, and White-throated

Eastern Towhee.

PHOTO BY JACK BARTHOLMAI

Eastern Towhee

Brown Thrasher

Vireos: Bell's and White-eyed

Warblers: Blue-winged Warbler, Chestnut-sided Warbler, Common Yellowthroat, Connecticut Warbler, Golden-winged Warbler, Mourning Warbler, Nashville Warbler, Prairie Warbler, Yellow-breasted Chat, and Yellow Warbler

American Woodcock

Wrens: including Bewick's, House, and Carolina

Cedar Waxwing

In addition, there are other species more closely associated with other habitats that will sometimes nest in shrubs and small trees in brushy areas. These include the Eastern Bluebird, Black-capped and Carolina Chickadee, Brown-headed Cowbird, Black- and Yellow-billed Cuckoo, Mourning Dove, Northern Flicker, American Kestrel, Great Crested Flycatcher, Eastern Kingbird, Northern Mockingbird, Eastern Screech-Owl, Ring-necked Pheasant, Tree Swallow, Tufted Titmouse, and Downy Woodpecker.

The shrubs selected in this chapter will provide every type of food required by various shrub-loving species of birds. The dogwoods, viburnums, blueberries, elderberries, cherries, and brambles produce berries, a favorite food for most shrubland birds. Available throughout the year, the berries of different shrubs ripen at various times from early summer to late fall, with the viburnum berries sometimes persisting until February. The nutrient content varies as well through the year. The summer berries are high in sugars (carbohydrates) and low in fats—perfect for parent birds busy searching for high-protein caterpillars to fatten their nestlings. The sweet berries are also needed by recently fledged young birds, who then need more carbs and less protein. The fall berries vary in their nutrients. Dogwood berries are high in fats—just what birds need to fatten up for their strenuous migrations. Viburnums, on the other hand, are generally lower in fats and consequently are less susceptible to early rot and can persist into the winter months, providing for birds during those lean times.

The other three shrub species offer seeds or nuts for birds. Hazelnuts ripen in summer and early fall, while St. John's wort and ninebark offer seeds that are available from July to as late as spring the following year.

Just as important, all these shrubs host many caterpillars and other insects, including bees for bee-eating birds, that offer vital food for many birds. These outstanding shrubs will attract birds to your yard and help them raise healthy young year after year, bringing joy and life to our yards.

DOGWOOD (*Cornus* species)

Dogwood Family, Cornaceae

Dogwoods are an outstanding group of native shrubs, with one or more species found in every state in the country. They have small, four-petaled flowers in flat-topped or rounded heads that form a platform for pollinating insects. The leaves have smooth margins and a distinctive vein pattern with side veins that branch out from the midvein and curve up in parallel to the leaf margins. All but pagoda dogwood have opposite leaves. The attractive berries arrive in bright white, blue, and red colors and are rich in fat to fuel the migration of the many ravenous birds that visit on their way south. And as fall progresses,

The pagoda dogwood has horizontal layers of branches resembling the architecture of a Japanese pagoda.

PHOTO BY MARIETTE NOWAK

the leaves turn a lovely burgundy. In winter, you can often distinguish species by stem color and branching structure. Many have upper twigs shaped like antlers, with flaring shoots of different lengths like the racks of deer. The pagoda dogwood is unmistakable with layers of branches like Japanese pagodas. Gray dogwood, commonly found in savannas, was highlighted in chapter 6. Four other common dogwoods found in shrubby habitats are featured here: pagoda (*Cornus alternifolia*), flowering (*C. florida*), blue-fruited or silky (*C. obliqua* or *amomum*), and red osier (*C. sericea*).

Value to Birds and Other Wildlife

See the description of gray dogwood in chapter 6 for an account of the many birds and insects that all dogwood shrubs support.

How to Grow

Most dogwood species prefer moist to wet soils, but many are tolerant of moderately dry soils. Full sun is recommended for the best flowering, fruiting, and twig color. The shrubby species spread by underground stems and benefit from occasional rejuvenation (cutting down to about six inches in winter or early spring). If desired, tree species like flowering dogwood can be selectively pruned in late fall or winter to shape and to remove diseased branches, but avoid doing so in early spring when wounds often bleed excessively. See the particular species descriptions for more details.

Propagation

Seeds and cuttings.

Companion Plants

The shrubby species have very dense growth and can be surrounded with herbaceous species that thrive in similar situations: wetland, woodland, or open habitats. Flowering dogwood, a tree, benefits from low ground covers, such as strawberry (*Fragaria* species), violet (*Viola* species), common oak sedge (*Carex pensylvanica*), or wild ginger (*Asarum canadense*), which don't obscure its beautiful form and also allow plenty of air circulation to prevent powdery mildew.

A Closer Look

Don't miss enjoying birds like American Robins and Gray Catbirds that visit dogwoods in fall. Their berries are sought after by many migrating birds who need their rich fat to bulk up before they leave to spend the winter in southern climes. I have to keep a close watch on my pagoda dogwood in fall; the berries barely ripen before they are devoured by birds.

A close examination of leaves of dogwoods is also in order as a way to verify that your shrubs are indeed dogwoods, despite the other varying characteristics of different species. Simply grab a leaf of any dogwood and gently pull apart the sections between the veins of the leaf a fraction of an inch. You'll see stringy white fibers that loosely hold the sections together (better seen with a hand lens). The elasticity of these fibers is unique to dogwoods and give the leaves strength.

Next, examine the leaves of dogwoods for rings or blister-like swellings. They are usually purplish with light centers and are made by the dogwood eyespot gall midge (*Parallelodiplosis subtruncata*). I found them on the gray dogwood in my yard, but they commonly occur on other species of dogwood as well. The midges are minute, mosquito-like flies that are seldom seen. They emerge from

The fat-rich berries of red osier dogwood provide fuel for migrating birds.
PHOTO BY ANNE MORETTI

the soil in spring and lay their eggs on the underside of the leaves. Tiny larvae grow inside galls and can sometimes be seen if you hold the gall to the light. In fall, the larvae drop to the soil and emerge as midges in spring. They rarely cause harm to dogwoods (mine are doing very well) and are usually kept in check by other insect parasites.

Pagoda Dogwood (*Cornus alternifolia*)

Deciduous Shrub or Small Tree
Size: Height 10–25 feet, width 6–15 feet
Color and Season: White flowers in spring to early summer; reddish-purple foliage in autumn
Light Requirement: Sun to shade
Soil: Moist to wet; rich, slightly acidic, well-drained soil
Native Habitat: Wet to moist woodlands
Native to: AL AR CT DE IA IL IN MA ME MD MI MN MO NC NH NJ OH PA RI SC TN VA VT WI WV; native but rare in FL LA NE
USDA Hardiness Zones: 3–7

This well-named species has a graceful horizontal layering of branches with distinctive upturned ends that resemble the architecture of a Japanese pagoda. It also is the only dogwood species with alternate, not opposite, branching, which gives it the Latin name *alternifolia*. It can be used as a specimen tree or in a shrub border. It is one of my favorite dogwood species, and I planted it as soon as I moved to a new home with a larger yard, where it thrives under the dappled shade of a white oak. It has large, two-inch-wide clusters of fragrant creamy-white flowers in early summer, and I make sure I don't miss the showy blue berries that follow before they are stripped by hungry birds. Just as attractive is its pleasing reddish-purple fall color. The unique tiered branching in winter provides interest throughout the year.

Pagoda dogwood has few insect problems, but I've had golden canker on mine, an increasingly common disease that affects only this species of dogwood. It is caused by a fungus (*Cryptodiaporthe cornia*) that turns twigs bright yellow or orange. The canker can spread to the rest of the plant and kill twigs, larger branches, and even the main trunk. The disease can be controlled by careful pruning of discolored twigs, disinfecting your pruners between cuts, and burning or burying the twigs. These practices have worked well for me, and my beautiful pagoda dogwood is thriving.

Flowering Dogwood (*Cornus florida*)

Deciduous Shrub or Small Tree
Size: Height 12–30 feet, width 8–15 feet
Color and Season: Ivory flowers with pink tints in spring; burgundy-red foliage in autumn
Light Requirement: Sun, partial shade
Soil: Moist to medium dry; slightly acid loam or clay loam
Native Habitat: Woodland edges or openings
Native to: AL AR CT DE FL GA IL IN KS KY LA MA MD ME MI MO MS NC NH NJ NY RI SC TN TX VA WV; native but rare in ME VT
USDA Hardiness Zones: 5–9

Flowering dogwood is native to southern midwestern and eastern states. Often considered the most spectacular of native flowering trees, it has showy flowers that set it apart from all other dogwoods. The petals are actually four bracts surrounding the true flowers—clusters of tiny quarter-inch yellowish-green flowers in the center of the bracts. The shiny crimson berries and alligator-hide pattern of its bark are also unique among dogwoods. Its burgundy-red fall foliage and graceful tiered branching add to its beauty through the seasons. Unfortunately, the introduced dogwood blight caused by aptly named anthracnose fungus (*Discula destructiva*) has killed many flowering dogwoods, but the disease is less likely to kill healthy trees. To reduce chances of its occurrence, plant it in open sunny locations, mulch well, and keep the soil moist during droughts.

Blue-Fruited Dogwood or Silky Dogwood (*Cornus obliqua; C. amomum*)

Deciduous Shrub
Size: Height 6–10 feet, width 3–6 feet
Color and Season: White flowers in late spring to early summer; purplish-red foliage in autumn
Light Requirement: Sun to partial shade
Soil: Wet, moist, medium; adaptable to variety of soils, tolerant of salt and compacted soils
Native Habitat: Wetlands from marshes to wet prairies, fens, and bogs
Native to: AR CT DE IA IL IN KS KY MA MD MI MN MO NE NH NY OH OK PA RI TX VT WI WV; native but rare in NJ SD TN VA
USDA Hardiness Zones: 4–8

This species grows in wetlands along with red osier dogwood but is not as common. It is our latest-flowering dogwood and has striking blue berries on green to reddish stems by late summer. The branches are purplish with grayish streaking, setting it apart from other dogwoods. It was recently separated from silky dogwood (*Cornus amomum*), but the taxonomy is still in flux and not accepted by all authorities. Blue-fruited dogwood is native to both the Midwest and northeastern states, while silky dogwood is primarily a northeastern species. Blue-fruited dogwood is a good choice for wet thickets and rain gardens and has been used for stream bank stabilization. It can also help choke out the invasives plants like glossy buckthorn (*Rhamnus frangula*), reed canary grass

A Cedar Waxwing feeds on the berries of the blue-fruited dogwood.
PHOTO BY JACK BARTHOLMAI

(*Phalaris arundinacea*), and non-native phragmites or common reed (*Phragmites australis*).

Red Osier Dogwood (*Cornus sericea; C. alba; C. stolonifera*)

Deciduous Shrub
Size: Height 3–10 feet, width 4–8 feet
Color and Season: White flowers in early spring; orange or red turning to purple foliage in autumn
Light Requirement: Sun, partial sun
Soil: Wet to medium
Native Habitat: Swamps, thickets
Native to: AK AZ CA CO CT DE IA ID IL IN KY MA MD ME MI MN MO MT ND NE NH NJ NM NV NY OH OR PA RI SD UT VT WA WV WI WY; native but rare in VA
USDA Hardiness Zones: 3–7

Red osier dogwood is a common shrub in wetlands and is widely planted in gardens for its bright red stems from fall to early spring—especially stunning

The red osier dogwood is widely planted for its bright red stems in fall and winter.
PHOTO BY MARIETTE NOWAK

against winter snows. It has the widest native range of any *Cornus* species and is found throughout most of the country, except the lower Midwest and Deep South. As a result of its wide range, there are many varieties and subspecies. A former name, *C. stolonifera*, refers to the stolons, trailing branches that form white roots where they contact the soil, enabling the shrub to spread and form dense thickets. Its specific Latin name, *sericea*, means silky, given for the fuzzy hairs found on the undersides of the leaves in some varieties. *Osier* is a French word meaning "willowy shoot," for the shrub's resemblance in winter to leafless willows. American Indians used its shredded bark as a substitute for tobacco, sometimes mixed with real tobacco. Prune often, as the young stems have the brightest red bark, although some varieties of this dogwood have yellow or brown stems. Plant in sunny areas in moist soils. Red osier dogwood is showy throughout the year with a profusion of white flowers in spring followed later in the season by bright white berries, lovely bronzy-purple fall leaves in fall, and ruby-red winter stems. I feel fortunate to have one in my yard to enjoy through the seasons.

HAZELNUT (*Corylus* species)

Hazelnut Family, Corylaceae

Hazelnuts are medium-sized shrubs that grow in savannas and open woods. They are the only nut-producing plant included among the shrub species featured in this chapter and add to the variety of foods available for shrubland birds. Although the nuts are not eaten by a great number of species, for some they are a preferred food and thus, importantly, help attract and support a greater

Hazelnut shrubs have attractive reddish-purple fall color.
PHOTO BY MARIETTE NOWAK

diversity of species in shrubby habitats. Moreover, hazelnuts are very nutritious, having 16 percent protein, 66 percent fat, and 18 percent carbohydrates. This proportion of proteins and fats is higher than that found in acorns.

Two species of hazelnut are native to North America, the American hazelnut (*Corylus americana*) and the beaked hazelnut (*C. cornuta*). The shrubs are often mentioned by the early land surveyors in the Midwest in the nineteenth century, who commonly wrote descriptions such as "scattering timber principally burr and white oak, hazel and hickory undergrowth." The easiest way to tell the two species apart is by their nuts. The American hazelnut has nuts enfolded in leaflike bracts ruffled at their tips and hidden among the foliage. The beaked hazelnut, as its name suggests, has nuts wrapped in tight beaks as much as an inch and a half long. The beaks have minute stiff hairs that can stick into and irritate the skin, like those of stinging nettle.

Hazelnuts have both male and female flowers on the same plant and are among the earliest flowers of the year to bloom, along with those of skunk cabbage. The male flowers are long drooping catkins, while the female flowers are tiny, exquisite, red starlike tufts at the ends of branches.

Value to Birds and Other Wildlife

The nutritious nuts are a preferred food of Ruffed Grouse, Ring-necked Pheasant, Hairy Woodpecker, and Blue Jay. Ruffed Grouse also eat hazelnut buds and catkins, important winter foods for them. Other birds also eat the nuts, including the Northern Bobwhite, Greater Prairie Chicken, Red-bellied Woodpecker, and Wild Turkey. Many birds also use the hazelnut thickets for cover and nesting sites. Most importantly, more than 130 species of butterfly and moth caterpillars find sustenance on the shrubs, making it among the top twenty woody plants supporting these protein-rich invertebrates, which are so important for birds and their nestlings.

Two of the moth caterpillars metamorphose into spectacular giant silk moths—the beautiful lime-green luna moth and the handsome polyphemus moth with its distinctive eyespot on each hindwing. Besides caterpillars, many other insects feast on various parts of hazelnut shrubs, including aphids, beetles, bugs, leafhoppers, sawfly larvae, and walking sticks. Several insects form galls, including the gall gnat (*Cecidomyia squamulicola*), which forms a gall in the form of a bulge at the base of the catkin. Another gall insect causes bent tips on the catkins.

Mammals that feed on the nuts include chipmunks, mice, and squirrels. Deer and rabbits browse on the leaves and twigs. Beaver use the stems both to eat and to build their dams. Like birds, mammals seek cover among the dense branches.

How to Grow

The shrubs produce the most nuts in the sun but are tolerant of partial shade as well as a wide range of soils. For the best nut production, plant in groups for better cross-pollination. Rather leggy shrubs with zigzagging twigs, they are best situated in natural habitats rather than more cultivated situations, and are perfect for informal hedges, thickets, and wilder areas of the yard.

Propagation

Nuts.

Companion Plants

Various sedges (*Carex* species), milkweeds (*Asclepias* species), sunflowers (*Helianthus* species), and goldenrods (*Solidago* species) are excellent companion plants, which often grow with hazelnut in the wild.

A Closer Look

In very early spring as we're all weary of winter, search out hazelnuts for their brilliant red, gemlike female flowers. The catkins may first catch your eye. Up to four inches long, they are beige in color and, when ripe, shed puffs of pollen grains if you tap them. You're also likely to find catkins with bulges at the base or bent tips caused by gall-forming insects. I cut these open but was unable to find the culprits. You may have better luck.

On the same branches as the catkins, search for the tiny female flowers. They are like buds with a crimson spray of stigmas, ready to receive pollen blowing in the breeze. Under a magnifying lens, they're spectacular! I've sought out these jewels every spring for years, first at the nature center where I worked and now in my own yard where I planted hazelnuts years ago and welcome their early blooms.

Autumn is the time to rummage among the hazelnut leaves to find the nuts, often hidden in the foliage, before hungry birds and squirrels devour them. The birds most likely to abscond with the nuts in your yard are Blue Jays, Hairy Woodpeckers, and Wild Turkeys. As I'm writing this in October, I'm also enjoying the lovely yellow-orange fall color of my trio of hazelnuts, another bonus of this interesting species. Don't miss enjoying hazelnut every season of the year.

American Hazelnut (*Corylus americana*)

Deciduous Shrub
Size: Height 5–12 feet, width 4–8 feet
Color and Season: Red flowers in very early spring; yellow-orange foliage in autumn
Light Requirement: Sun to partial sun
Soil: Medium, medium dry, medium wet; variety of soil types—loamy, sandy, or rocky
Native Habitat: Savannas, open woodlands
Native to: AL AR CT DE GA IA IL IN KS KY MA ME MD MN MO MS NC ND NE NH NJ NY OH OK PA RI SC SD TN VA WI WV; native but rare in LA VT
USDA Hardiness Zone: 4–9

This is a somewhat unkempt shrub, best used in naturalized areas. I grow three of them in a shrubby edge between a pocket prairie and wooded area, where they're free to sprawl among other shrubs. Hazelnuts can easily be shaded out by other shrubs, however, so I prune back the taller shrubs each year to make sure my hazelnuts have sufficient sun. This species does well in a somewhat drier, sunnier site than does its cousin the beaked hazelnut.

Beaked Hazelnut (*Corylus cornuta*)

Deciduous Shrub
Size: Height 5–14 feet, width 4–8 feet
Color and Season: Red flowers in very early spring; yellow-orange foliage in autumn

Light Requirement: Sun to partial sun
Soil: Medium, moist; acidic, rich, well-drained soil; tolerates average garden soils but not heavy clay
Native Habitat: Open woodlands, savannas
Native to: AL CA CO CT GA IA ID MA ME MI NC ND NH NJ OR PA RI SC SD TN VA VT WA WI WV; native but rare in DE IL MD WY; extirpated in OH
USDA Hardiness Zones: 4–8

This hazelnut prefers somewhat shadier and more moist sites than does the American hazelnut. It does well in woodland openings or shadier spots in savannas but does not tolerate completely open, sunny sites. The stiff irritating hairs on the beaks that wrap the nuts have likely evolved to protect the nuts from predation by animals.

ST. JOHN'S WORT (*Hypericum* species)

St. John's Wort Family, Clusiaceae

There are several shrub species in this genus and all are mound-shaped plants with dense branching and marvelous flower displays. The five-petaled flowers are centered with a bushy tuft of stamens loaded with bright yellow pollen. They mantle the mounds with blooms for two or more months, adding cheerful, long-lasting color in midsummer.

The common name, St. John's wort, was given because it often begins blooming around June 24, the feast day of Saint John the Baptist. The word *wort* was used historically for plants valued for their medicinal properties or for food; an herbaceous *Hypericum* species has been used medicinally, particularly to treat mild depression.

By fall, the flowers fade and form chocolate-brown seed capsules that persist through the winter, adding winter interest and food for birds. The fall color ranges from bright red in plants in full sun to yellowish for those in more shade but can vary season to season. The peeling bark on older stems, in cinnamon-brown to purple colors, is also attractive.

St. John's wort is mantled in bright yellow blossoms for two or more months in summer.

PHOTO BY MARIETTE NOWAK

Two species, Kalm's St. John's wort (*H. kalmianum*) and shrubby St. John's wort (*H. prolificum*), will be featured here. Both species do equally well in my yard and are among my favorite shrubs. There are several other woody species of *Hypericum*, native primarily to far eastern or southeastern states.

Value to Birds and Other Wildlife

The seeds of shrubby St. John's wort, enclosed in small brown capsules, are an abundant source of seeds for birds. The seeds are important as winter food for the American Tree Sparrow, Dark-eyed Junco, Northern Bobwhite, Ring-necked Pheasant, Ruffed Grouse, and some waterbirds. The dense branching also provides excellent cover for birds.

Butterflies will occasionally stop for nectar, only to be disappointed, as the flowers of the shrubs have no nectar; instead, the plants have put all their energy into producing pollen for bees. Eighteen species of caterpillars feed on the shrubs, adding an important protein source for birds. Caterpillars of the gray hairstreak butterfly feed on the seed capsules. The caterpillar of the wavy-lined emerald moth prefers to munch on the flowers. Many other moth caterpillars feed on various other parts of the plants, including those of the gray half-spot, which feed on the foliage. Other insects that feed on various parts of the shrubs include aphids, beetles, bugs, leafhoppers, and weevils. Mammals avoid the shrub because its foliage has a chemical that irritates their skin and intestines.

One of the greatest contributions of these shrubs is the prodigious amounts of pollen they offer bees and other pollinators. A true magnet for bees, St. John's wort is the best "bee" plant in my yard. On sunny, warm days, you can hear a steady buzz as the bees dance among the blossoms. A starburst spray of stamens in the midst of the petals carry the precious pollen that the bees seek for their larvae. The bees emerge from the flowers with their tiny leg sacks filled with the golden treasure and bodies powdered with the pollen, some of which will be deposited at the next flower they visit. Bumblebees are the most important pollinators of the shrub, but other insects, including halictid bees and syrphid flies, also visit, although they are not as effective pollinators.

The visiting bees attract several species of birds that feed on them, particularly the Summer Tanager. It specializes in catching bees and wasps, although it eats many other insects as well. The tanager will catch the bee in midair and take it to a tree and beat it against a branch until dead. Then it rubs the bee against the branch to squeeze out the stinger before eating it. The closely related Scarlet Tanager occasionally eats bees as well, as do Eastern Kingbirds, Purple Martins, and woodpeckers.

The major reason bees, particularly top pollinators like our wild bumblebees, are important for birds is that they pollinate most of the plants that provide berries, nuts, and seeds for birds, as highlighted by Rich Hatfield of the Xerces Society. More than 85 percent of flowering plants require an animal, primarily bees and other insects, for pollination. They are vital for the pollination of our own food crops as well as the "food crops" of birds and wildlife. Because of their pollination services, experts consider bees to be "keystone species" playing a key role in the ecosystems and enabling other species to survive. So it's very worrisome that bees, both the non-native honey bees and wild bees, are declining throughout North America. Bee plants, like St. John's wort, are vital for both bees and birds.

How to Grow

St. John's wort shrubs are ultra-hardy, adaptable to soils ranging from sandy and rocky to heavy clay. Young plants need to be watered until well established.

They have few diseases or insect problems and are tolerant of the soil toxins of walnut trees. Use them as specimen plants or in masses for a showy shrub garden. Having shallow roots, they are easy to transplant and I've given many small shrubs to friends. They spread readily by seed, so if you want to contain them, clip off the seed heads in late fall or winter (although this will deprive birds of the seeds). To ensure dense branching, prune stems down about a fourth, not by shearing off but by cutting each branch down to a lower bud. I failed to prune mine and they became very rangy and unkempt, until I began to do so.

Propagation

Seeds and cuttings.

Companion Plants

Since St. John's wort is in full bloom in midsummer, consider surrounding them with species that bloom earlier or later in the year for all-season bloom. In spring, my shrubs are surrounded with wild strawberry (*Fragaria virginiana*) and Canada anemone (*Anemone canadensis*). For late summer bloom, asters and goldenrods are excellent choices.

A Closer Look

Enjoy the buzz and bustle of the many bees attracted to this shrub. How many species can you find? If you, like me, are not an expert on bees, simply count the different kinds and celebrate the diversity you are bringing to your yard.

St. John's wort blossoms abuzz with bees collecting pollen.
PHOTO BY MARIETTE NOWAK

The bees are literally too busy gathering pollen to bother you and won't sting unless disturbed. You may also be lucky to see tanagers and kingbirds, zapping up the bees. A careful search may uncover caterpillars of moths and butterflies, usually very skilled at concealing themselves. Later in the year, juncos flock to the shrubs to feed on the dark seed capsules, one of their mainstays throughout the winter months in the Upper Midwest where I live.

Kalm's St. John's Wort (*Hypericum kalmianum*)

Deciduous Shrub
Size: Height 2–3 feet, width 2–3 feet
Color and Season: Bright yellow blooms in summer; yellow to red foliage in autumn
Light Requirement: Sun
Soil: Moist to dry; tolerates many soils
Native Habitat: Prairies
Native to: MI WI; native but rare in IL IN OH; extirpated in NY
USDA Hardiness Zones: 4–7

This species is named after Peter Kalm, a student of Linnaeus, who discovered the shrub in North America in the eighteenth century. As early as 1783, the shrub was being offered for sale in America's first plant catalog published by Bartram Gardens.

Shrubby St. John's Wort (*Hypericum prolificum*)

Deciduous Shrub
Size: Height 3 to 6 feet, width 3–6 feet
Color and Season: Bright yellow blooms in summer; yellow to red foliage in autumn
Light Requirement: Sun, part sun
Soil: Moist to medium dry; rocky barren soils; tolerant of salt and compaction
Native Habitat: Thickets, swamp borders, open wood
Native to: AL AR GA IA IL IN KY LA MD MI MO MS NC OH OK PA SC TN TX VA WI WV; native but rare in DE FL NJ NY; questionable native presence MN
USDA Hardiness Zones: 3–8

St. John's wort has bright red fall color some years, but color can vary season to season.
PHOTO BY MARIETTE NOWAK

Shrubby St. John's wort is hard to distinguish from Kalm's St. John's wort. Experts distinguish the two by their seed capsules. Shrubby is said to have three-celled seed capsules with three style tips, while Kalm's has five-celled seed capsules with five style tips. However, I have two specimens labeled shrubby St. John's wort from the same nursery, and one has three-celled capsules and the other has four-celled capsules (a hybrid?). One is in full blossom by June 30; the other begins to blossom in July. The species name, *prolificum* (spreading), is quite apt, as these shrubs spread readily by seed in my yard and are forming a lovely shrub island, without any effort on my part.

NINEBARK (*Physocarpus opulifolius*)

Rose Family, Rosaceae
Deciduous Shrub
Size: Height 6–12 feet, width 6–10 feet
Color and Season: White flowers in spring; yellow to orange foliage in autumn
Light Requirement: Sun, partial sun
Soil: Moist to dry; acidic, alkaline, heavy clay; tolerates salt

Native Habitat: Shores of streams, lakes, and ponds; moist and dry prairies; open riparian woodlands
Native to: AL AR CO GA IA IL IN KY MD MI MN MO NC NJ NY OH PA SC TN VA WI WV; native but rare in FL KS ND NE OK SD; extirpated in DE
USDA Hardiness Zones: 2–8

With arching cascades of blossoms in spring, ninebark is an excellent native substitute for the closely related bridal wreath (*Spirea prunifolia*), a shrub from China, Korea, and Japan. A multistemmed shrub, ninebark has broad, arching branches that give it a mound-like shape. Smaller side stems bear the spring flowers and summer fruits. Its flowers are typical rose family flowers with five petals centered with a tuft of stamens. They are borne in two-inch-wide clusters that bloom for two to three weeks.

Its species name, *opulifolius*, was given because its lobed leaves resemble those of cranberrybush viburnum (*Viburnum opulus*), whose species name is similar (*opulus*). Ninebark is as showy in fall fruit as it is in spring flower. Each of its flowers produces three to five seedpods—inflated capsules called follicles. Ninebark's genus name, *Physocarpus*, refers to these inflated seedpods and comes from the Greek words *physo* (bladder or bellows) and *carpus* (fruit). Seedpods are about a third inch long. Initially greenish red, they turn a beautiful pinkish red before becoming brown in late fall.

The ninebark shrub is a favorite nesting site for birds like this Yellow Warbler.
PHOTO BY JACK BARTHOLMAI

This shrub offers winter interest as well, when its peeling bark in many shades of beige, brown, and cinnamon are revealed. The bark exfoliates in multiple layers, hence the common name, ninebark.

Value to Birds and Other Wildlife

Ninebark offers a variety of perks to birds—food, shelter, nesting material, nest sites, and a place to stash seeds. Hummingbirds visit for its nectar. Waterfowl, marsh birds, shorebirds, and upland game birds relish its seeds. A wide range of caterpillars and other insects are also attracted to ninebark and offer nourishment for both adult and nestling birds. Ground birds like turkeys, quail, and pheasant find shelter in its closely spaced stems. Chickadees cache food morsels within its shredding bark, and a variety of birds gather strips of bark for nesting material. The shrub is a popular nesting site, where nests are safely

Ninebark has showy clusters of blossoms with yellow centers and tufts of white stamens.
PHOTO BY ANNE MORETTI

hidden in its leafy interior. I have personally discovered more songbird nests, including those of goldfinches, sparrows, and warblers, in ninebark than in any other shrub.

The flowers provide both nectar and pollen to many bees, bugs, beetles, wasps, flies, moths, and butterflies. The nectar at the base of the stamens is easily accessed by the insects. The fifteen stamens, when mature and loaded with pollen, extend well out from the flower petals. Visiting insects can't avoid being covered in pollen as they feed and so help spread it from flower to flower, ensuring pollination. The flowers are a magnet for red admiral butterflies—one observer saw twenty-five of them on the blossoms of one shrub! Other butterflies fond of the nectar include the American lady, monarch, red-spotted purple, spring azure, and viceroy. The handsome day-flying ctenucha moth, with its two-inch wingspan, also visits for nectar.

As larvae, many of these and other insects feed on the leaves, stems, and other plant parts. The list is long and includes aphids, witch hazel borer, plant bugs, and more than forty species of butterfly and moth caterpillars. Three of the moth caterpillars have a strong preference for ninebark and similar shrub species, including those of the dimorphic eulithus, raspberry leafroller, and the appropriately named ninebark pygmy moth. Other moth caterpillars that feed on ninebark as well as a variety of other plants are those of the dark-spotted palthis, the white spring, and the unicorn.

An insect of particular interest is the beautiful ninebark leaf beetle (*Calligrapha spiraea*). It acquired its species name, *spiraea*, because ninebark was then lumped in the same genus as the shrub *spirea*. Ninebark has since been put in a separate genus, but you'll find that some references still use the old name for the beetle, calling it the spirea leaf beetle. The beetle has a rounded shell-like exoskeleton with colorful black marks and dots on its reddish-orange back. The markings resemble calligraphy and are a distinctive feature of this and other beetles in the aptly named *Calligrapha* genus. Unfortunately, horticulturists have developed a popular cultivar of ninebark with purple leaves instead of the natural green leaves. The ninebark beetle is repelled by these purple leaves, likely due to chemical changes in the leaf caused by the color change. Research has found that changes in leaf or petal color in plants often deter insect feeding. For this reason, it's best to stick with straight natives and avoid the use of cultivars of native plants, often called nativars, such as the purple ninebark. Straight native plants, which have coevolved with insects over millions of years, are usually the best choice to support our native wildlife.

How to Grow

One of the hardiest of shrubs, ninebark is adaptable to nearly every soil condition, wet or dry, and will thrive in sun to part shade. It is resistant to drought, heat, and soil compaction and has no serious insect or disease problems. It is an excellent choice for a foundation shrub, since it is so easy to maintain. It can also be used in hedges, shrub borders, and wetland restorations. Prune and shape as needed; alternatively, cut to the ground every few years. I've planted this beautiful shrub—one of my favorites—at my former and current homes.

Propagation

Easy from cuttings and seed.

Companion Plants

Whether you plant it in moist or dry soils, surround the shrub with native sedges and wildflowers that won't obscure ninebark's cascades of blossoms in spring or its beautiful seedpods in early fall. Some wildflowers that fill the bill are showy goldenrod (*Solidago speciosa*), sky blue aster (*Symphyotrichum oolentangiense*), and wild bergamot (*Monarda fistulosa*).

Pinkish-red seedpods decorate ninebark in late summer.
PHOTO BY MARIETTE NOWAK

A Closer Look

You're sure to admire the mounds of blossoms in spring, but don't miss ninebark's beauty other seasons of the year, particularly in late summer as its seedpods form. The abundant pods, inflated like tiny balloons in rich pinkish-red hues, resemble ornaments strung along its branches. I picked some of these seedpods in fall when they were already brown and dried them on my desk. They opened overnight and resembled miniature bouquets of dried flowers. Inside you'll find tiny, tear-shaped tan seeds. Fall or winter are also excellent seasons to check for bird nests that may have been hidden in the dense foliage during the summer.

CHERRY SHRUBS (*Prunus* species)

Rose Family, Rosaceae

Plants in the *Prunus* genus are among the most valuable for wildlife, supporting an incredible diversity of insects and producing fruits that nourish nearly fifty species of birds. Shrubs in the genus include both cherries and plums. Their stony fruits, called drupes, are greatly relished by birds. The white flowers, though relatively small, are five-petaled beauties with a central tuft of stamens displayed lavishly on the plants. The flowers of most species silhouette the

Chokecherrry's blossoms gracefully drape from its branches.

PHOTO BY MARIETTE NOWAK

shape of the plants in a gauzy veil of blossoms. But chokecherry (*P. virginiana*) and the black cherry tree (*P. serotina*) have a different flowering arrangement—clusters of flowers gracefully drape from their branches in abundance, like tinsel on a Christmas tree.

The plums are in their glory in spring, commonly coloring our roads with thickets of white blossoms. Their blackish bark is easily distinguished, even at a distance, from the gray bark of serviceberries (*Amelanchier* species) that blossom at the same time. The most widespread plum is simply called wild plum (*P. americana*) and is found in nearly the entire country with the exception of some far western states. Canada plum (*P. nigra*) is mostly a Canadian species as its name implies, while beach plum (*P. maritima*) has a very small northeastern range. Although I love wild plum and look forward to its extravagant blossoming every spring, it is not a "bird shrub." Its larger, gumdrop-sized fruits are too large for birds and are eaten by few mammals except for fox. Some experts believe that the fruit of plums, like those of the Kentucky coffee tree and locust trees, was eaten and dispersed by mammals that are now extinct.

Two species of cherry shrubs, with smaller bird-sized fruits of about a third inch in diameter, are commonly available for landscaping—chokecherry (*P. virginiana*) and sand cherry (*P. pumila*). Both will be highlighted in this chapter. Two cherry trees are featured in chapter 8.

Value to Birds and Other Wildlife

Cherry fruits, except for plums, are a favorite food for thirty-two birds, including the American Crow, American Robin, Baltimore Oriole, Bicknell's Thrush, Blue Jay, Brown Thrasher, Cedar Waxwing, Eastern Bluebird, Eastern Kingbird, *European Starling*, Evening Grosbeak, Gray Catbird, Gray-cheeked Thrush, Hermit Thrush, Northern Bobwhite, Northern Cardinal, Northern Flicker, Northern Mockingbird, Orchard Oriole, Pileated Woodpecker, Red-bellied Woodpecker, Red-eyed Vireo, Red-headed Woodpecker, Ring-necked Pheasant, Rose-breasted Grosbeak, Ruffed Grouse, Summer Tanager, Swainson's Thrush, Veery, Warbling Vireo, Wood Thrush, and Yellow-bellied Sapsucker. Another fifteen species grab an occasional bite, including the American Goldfinch, Common Grackle, Downy Woodpecker, Eastern Towhee, Great Crested Flycatcher, Greater Prairie Chicken,

Hairy Woodpecker, *Monk Parakeet*, Pine Grosbeak, Red-winged Blackbird, Scarlet Tanager, Song Sparrow, White-crowned Sparrow, White-throated Sparrow, and Wild Turkey. In addition to eating the fruit, the Yellow-bellied Sapsucker, as its name suggests, feeds on the sap. American Robins, Baltimore and Orchard Orioles, and Eastern Kingbirds are known to nest in the cherries.

Many bees and flies, plus some butterflies and skippers, visit the flowers of cherry shrubs and trees for pollen and nectar. Most important for birds are the caterpillars that feed on the leaves and other parts of cherries, totaling more than 450 lepidoptera, all of which provide vital food for birds. The caterpillars include the coral hairstreak, red-spotted purple and tiger swallowtail butterflies, and ten species of giant silk moths, such as cecropia moth, imperial moth, io, and polyphemus moths. Some of these caterpillars and other insects specialize in cherries, as suggested by their names, including the caterpillars of

Northern Cardinals are among the many birds that feed on chokecherry's fruits.
PHOTO BY JACK BARTHOLMAI

cherry blotch miner moth and cherry leaf roller moth, chokecherry aphid, chokecherry fruit fly, and chokecherry leaf beetle.

How to Grow

The shrub cherries are self-fertile, but for lush flowering and fruiting, plant two to three shrubs.

Black knot, a common disease seen on cherries, is caused by a native fungal disease that forms black tarry swellings on the branches. The fungal spores are windborne, making it hard to prevent the disease. The best approach is to cut off the knots and burn them (don't add to your compost pile). Despite the ubiquity of black knot, it rarely causes much harm to cherries, possibly because native insects prey on or parasitize it.

Propagation

Seeds, wood and root cuttings.

Companion Plants

Since cherries grow in a wide range of soils from dry to moist, choose herbaceous species suited to the soil in which they are to be planted. In dry conditions, cherries are known to grow with common oak sedge (*Carex pensylvanica*), common spiderwort (*Tradescantia ohiensis*), and wild lupine (*Lupinus perennis*). In moist soils, cherries are found with blue-stemmed goldenrod (*Solidago caesia*), feathery false Solomon's seal (*Smilacina racemosa*), and spring beauty (*Claytonia virginica*).

A Closer Look

Cherry shrubs are some of the first woody plants to flower and fruit in spring. Watch for many fruit-eating birds, eager to feast on the first berries of the season. All our thrushes and woodpeckers are particularly fond of the cherries and may be among the first birds to arrive. Also take time to smell the crushed cherry leaves and twigs, which to me have a sweetish, sometimes stale odor. It has been described by others as a strong, bitter-almond smell and is caused by cyanide, a poison that is also found in cherry pits.

Sand Cherry (*Prunus pumila*)

Perennial Wildflower
Size: Height 1–3 feet, width 3–6 feet
Color and Season: White flowers in spring; yellow to red-orange foliage in autumn
Light Requirement: Sun
Soil: Moist to dry; sandy soils
Native Habitat: Sandy prairies, savannas, and beaches
Native to: CO CT DE IN IL KS MA ME MI MN ND NE NY SD UT VI WI WV; native but rare in IA KY NJ MD MT OH PA RI TN VA WV; extirpated in NC
USDA Hardiness Zones: 3–7

This cherry has been called the beauty of the dunes with its showy clusters of flowers in spring. There are many varieties of sand cherry throughout its native range, which can be a challenge to distinguish. The species name *pumila* (dwarf) refers to the shorter varieties. Sand cherry is a hardy plant adaptable to extremes of temperature and precipitation and grows well in sunny locations on rocky, loamy, and clay soils. It's an excellent choice for restorations and native landscapes.

Choke Cherry (*Prunus virginiana*)

Shrub/Small Tree
Size: Height 35–50 feet, width 20–35 feet
Color and Season: White flowers in spring; golden-yellow to orange foliage in autumn
Light Requirement: Sun; partial shade
Soil: Moist to dry; sand, clay
Native Habitat: Woodlands, savannas, thickets
Native to: All states except AL FL HI LA MS SC; native but rare in GA TN
USDA Hardiness Zones: 2–7

Choke cherry is a beautiful shrub for naturalized settings. It's a bit on the wild side for more conventional landscapes due to its suckering habits and tangle of branches. But in my yard, it forms a very attractive grouping on a lower slope. Its almond-scented flowers are followed by red berries that ripen to dark bluish-black berries beloved by birds, and the shrub has lovely yellowish-orange foliage in fall. Use it as specimen shrub, hedge, or woodland edge planting.

BRAMBLE BERRY (*Rubus* species)

Rose Family, Rosaceae

The Latin word *Rubus* translates to "bramble," meaning bushes with thorns, and today these berry bushes are also commonly referred to as brambles. Among the many *Rubus* species are the big three—blackberry (*R. allegheniensis*), black raspberry (*R. occidentalis*), and red raspberry (*R. idaeus*), all of which occur in many states across the country. Two other, less common brambles are purple-flowered raspberry (*R. odoratus*) and thimbleberry (*R. parviflorus*). All five are featured here.

When purchasing your berry bushes, it's best to go to a reliable native plant nursery. Be sure to avoid non-native species, especially Himalayan blackberry (*R. discolor*) and wineberry (*R. phoeniolasius*), which have become extremely invasive in the Pacific Northwest and along the Atlantic coast, respectively.

Value to Birds and Other Wildlife

Brambles offer safe havens and veritable grocery stores for a multitude of birds and insects. The fruits have very high value for many upland game birds

Brambles, like this blackberry, are not only beautiful in flower but highly valuable to birds.
PHOTO BY MARIETTE NOWAK

and songbirds. They are a preferred food for twenty-five species of birds: Cedar Waxwing, Northern Cardinal, Rose-breasted Grosbeak, Scarlet Tanager, Summer Tanager, Fish Crow, Blue Jay, Fox Sparrow, Eastern Towhee, Pine Grosbeak, Baltimore Oriole, Orchard Oriole, Common Grackle, Gray Catbird, Northern Mockingbird, Brown Thrasher, Tufted Titmouse, Yellow-breasted Chat, Bobwhite Quail, Wild Turkey, Ring-necked Pheasant, Greater Prairie Chicken, Veery, Wood Thrush, and American Robin. Other birds that feed on the fruits include Lesser Scaup, Blue Grosbeak, Indigo Bunting, American Crow, Evening Grosbeak, Red-winged Blackbird, Rusty Blackbird, Northern Flicker, Red-bellied Woodpecker, Red-headed Woodpecker, Hairy Woodpecker, American Woodcock, Hermit Thrush, Gray-cheeked Thrush, Swainson's Thrush, Eastern Bluebird, Alder Flycatcher, Willow Flycatcher, Great Crested Flycatcher, Eastern Phoebe, White-eyed Vireo, Eastern Kingbird, Red-eyed Vireo, Henslow's Sparrow, Song Sparrow, Field Sparrow, and White-throated Sparrow. Ruffed Grouse have a preference for the berries and eat the buds as well.

Many of the birds mentioned here are also known to nest in brambles, including Willow and Alder Flycatchers, Northern Mockingbird, Gray Catbirds, Brown Thrasher, White-eyed Vireo, Yellow-breasted Chat, Northern Cardinal, Blue Grosbeak, Indigo Bunting, and Field and Song Sparrows.

Insects abound in brambles and provide valuable protein for birds and their nestlings. The flowers' nectar and pollen attract bees, wasps, flies, ants, butterflies, beetles, bugs, aphids, skippers, and beetles. A host of other insects feed on the roots, leaves, fruits, or canes of brambles. More than 160 butterfly and moth caterpillars feed on the shrubs, as well as beetles, plant bugs, stink bugs, aphids, leafhoppers, treehoppers, katydids, walking sticks, thrips, and the larvae of sawflies. Many mammals also feed on the fruits, canes, leaves, and seeds of brambles. Deer, mice, opossum, raccoon, fox, chipmunks, rabbits, squirrels, moose, elk, and bears—all like to munch on brambles despite the thorns.

How to Grow

All brambles prefer sun and well-drained soil. There are specific guidelines for growing various cultivated species of raspberries and blackberries, but in the wild, the rambunctious brambles usually grow in abandon, spread readily without care, and are perfect for the less groomed, untamed corners of your yard, as they can become aggressive in good conditions. Straggly wild plants or transplants to your yard will produce more fruit if other plants are cleared from within two feet of each cane (stem) and then mulched around but not touching the stem. Remove dead, diseased, or thinner canes to produce more attractive and healthier bushes.

When you gather berries, you may occasionally see a white spot on some of the small balls (drupelets) on the berry that surround the seed. This is called white drupelet syndrome and is caused by sunscald and sometimes by high temperatures and even wind. The berries are still fully safe to eat but are usually unacceptable in commercial markets because they're considered unsightly. Sometimes the white spots are caused by spider mites or stink bugs, making the berry tan in color or white in irregular patterns. Check the leaves to detect these critters. If the bottom of the leaves are dusty, mites are the problem. Stink bugs are easier to spot and will appear as black specks on the leaves or much larger as they grow. Natural predators like lacewings and ladybugs usually keep these pests in check. It's best not to eat whitish berries caused by insect feeding.

Propagation

Root cuttings, layering branch tips, transplanting suckers, seeds.

Companion Plants

I like to grow common oak sedge (*Carex pensylvanica*) around my berry bushes in place of mulch. This hardy, grasslike sedge helps retain soil moisture, reduces weeds, and makes a lawn-like surface on which to stand when picking berries.

A Closer Look

You're sure to see many birds gorging on the abundant raspberries and blackberries in your yard. Among them, watch for three major fruit eaters in the Midwest: the mimic thrushes, including the Brown Thrasher, Gray Catbird, and Northern Mockingbird. All three feed on a variety of berries, but the berries of *Rubus* species are a particular favorite of the thrasher. Not only do mimic thrushes make berries a major part of their diets, but they also nest in dense shrubs like brambles. In fact, the shrubs are one of the most frequently used

The Brown Thrasher, along with catbirds and mockingbirds, is particularly fond of the berries of brambles.

PHOTO BY JACK BARTHOLMAI

nest sites of the catbird. All three birds are favorites of many bird-watchers. It's not because they have showy coloration; many warblers have far more showy plumages. No, it's their singing that has made them famous. Each has its own repertoire and can easily be identified by its songs. The catbird doesn't repeat but sometimes meows like a cat, the thrasher sings in pairs, and the mockingbird sings each phrase three or more times.

Common Blackberry (*Rubus allegheniensis*)

Deciduous Shrub
Size: Height 3–8 feet, width 4–8 feet
Color and Season: White flowers in late spring / early summer; red foliage in autumn
Light Requirement: Sun, part shade
Soil: Moist; prefers rich fertile soil but tolerates clay loam or rocky soils
Native Habitat: Open woods, prairies, roadsides
Native to: AR CT DE GA IA IL IN KS KY MA MD ME MI MN MO NC NE NH NJ OH OK PA RI SC TN VA VT WI WV; native but rare in AL
USDA Hardiness Zones: 3–8

Common blackberries have large showy flowers, up to an inch across, and bloom for about a month. The fruits are usually larger than those of raspberries, generally three-quarters of an inch long and a third inch across, although their size varies with moisture levels. Green at first, they become red, then black when fully ripe. It can be difficult to tell different blackberry species apart, but this species is the most common and has the largest berries. Blackberries in fall, with their rich red foliage glowing in the sun, were described by Aldo Leopold in his classic book, *A Sand County Almanac*, as "red lanterns" that guided his way through bottomland woods.

Red Raspberry (*Rubus idaeus var. strigosus*)

Deciduous Shrub
Size: Height 2–6 feet, width 3–5 feet
Color and Season: Pink to purple flowers in spring to summer; maroon-red foliage in autumn
Light Requirement: Sun, part shade
Soil: Medium moisture; rich, slightly acidic, well drained
Native Habitat: Open woods, ravines, bluffs
Native to: All states except AL AR FL HI KS KY LA MS SC TX
USDA Hardiness Zones: 4–8

This is the most widely spread *Rubus* species in the United States. It has small white flowers and very tasty red berries. Red raspberries are grown not only for their fruit but sometimes for a foliage accent as well. There are two main varieties, the native American red raspberry (*R. idaeus var. strigosus*) and the European raspberry (*R. idaeus var. idaeus*). Commercially grown raspberries are mostly cultivars of these varieties or crosses between the two.

Black Raspberry (*Rubus occidentalis*)

Deciduous Shrub
Size: Height 3–6 feet, width 4–8 feet
Color and Season: White flowers in late spring / early summer; bluish-green to yellow foliage in autumn

Light Requirement: Partial sun
Soil: Moist; rich loam; tolerant of compaction, salt
Native Habitat: Open woods, thickets, savannas
Native to: AL AR CO CT DE GA IA IL IN KS KY LA MA MD ME MI MN MO MS NC ND NE NH NJ OH OK PA RI SC SD TN VA VT WI WV
USDA Hardiness Zones: 4–8

This ubiquitous species has flowers that are smaller than those of blackberries, about a half inch across. They bloom for two to three weeks, forming white berries that soon turn red, then dark purple when fully ripe, and have a delicious sweet-tart flavor. An identifying feature is the whitish bloom, termed *glaucous* by botanists, on the young canes that can be rubbed off.

Purple-Flowered Raspberry (*Rubus odoratus*)

Deciduous Shrub
Size: Height 3–6 feet, width 4–8 feet
Color and Season: Pink to purple flowers in summer; yellow foliage in autumn
Light Requirement: Sun, part shade, shade
Soil: Moist, medium dry; good garden soil; tolerant of compaction, salt
Native Habitat: Woodland/forest borders and clearings
Native to: AL CT DE KY MA MD ME MI NC NH NJ NY OH PA RI SC TN VA VT WI WV; native but rare in IL IN
USDA Hardiness Zones: 3–8

This beautiful fragrant shrub is native to five midwestern states but is primarily found in the East. It is prized for its purple flowers, which bloom profusely throughout the summer season when grown in sunny locations and host a bevy of pollinators. Unlike most *Rubus* species, this species is thornless. Its large berries lack sweetness but are highly sought after by fruit-eating birds (most of which have no or a limited sense of taste). A single shrub is somewhat self-fertile, but it's best to have two or three of the shrubs to produce abundant fruit.

Thimbleberry (*Rubus parviflorus*)

Deciduous Shrub
Size: Height 4–8 feet, width 4–8 feet
Color and Season: White flowers in early to late summer; yellow foliage in autumn
Light Requirement: Full sun, partial shade
Soil: Dry to moist; adaptable but prefers fertile, well-drained soils
Native Habitat: Woodland/forest edges
Native to: AZ CA CO ID MI MN MT NM NY OR SD UT WA WI WY
USDA Hardiness Zones: 3–10

Thimbleberry is native to four midwestern states but is mainly a western species. This beautiful shrub has showy, white flowers that attract many pollinators. The delicious berries are irresistible to birds and wildlife. A single shrub is somewhat self-fertile, but it's best to have two or three of the shrubs to produce abundant fruit. Avoid cutting first-year canes; unlike most other raspberries, older two- to three-year-old canes produce the most fruit.

WILLOW (*Salix* species)

Willow Family, Salicaceae

See pussy willow (*Salix discolor*) in chapter 4 for a general introduction to willows and a description of their value for birds. The two additional shrub

The American Redstart Warbler often breeds in willow thickets.
PHOTO BY JACK BARTHOLMAI

willows that are commercially available and suitable for landscaping are Bebb's willow (*S. bebbiana*) and prairie willow (*S. humilis*). Companion plants for these two species are given in the following descriptions. All three willow shrubs (as well as willow trees) are among the most important plants for birds, offering a bounty of seeds, nesting materials, nourishing insects, and nest sites. Willow shrubs can often be easily identified in the field by their galls, particularly the willow pine cone gall.

Bebb's Willow (*Salix bebbiana*)

Deciduous Shrub or Small Tree
Size: Height 20–35 feet, width 10–15 feet
Color and Season: Yellow-tipped stamens on male catkins and greenish female catkins in spring; insignificant foliage color in autumn
Light Requirement: Sun
Soil: Wet to moist; rich soils
Native Habitat: Wet prairies, marshes, riverbanks
Native to: CO CT IA ID IL IN MA ME MI MN MT ND NH NJ NM NV NY OH OR PA RI SD VT WA WI WY; rare in AZ CA NE; extirpated in DE MD
USDA Hardiness Zones: 3–7

This willow thrives in a variety of moist habitats where it grows either as a multistemmed shrub or a small tree with a single trunk. It is often found with other willows and sometimes hybridizes with them. Bebb's willow establishes easily and is used to stabilize shorelines and for buffer plantings. The common name refers to Michael Schuck Bebb, an American specialist in willows. This species is one of the most important "diamond" willows, which have diamond-shaped patterns on their trunks caused by a fungus. An attractive diamond-shaped contrast between the white sapwood and reddish-brown heartwood becomes prominent when the stems are carved, and they are used for candle-holders, canes, furniture, and lampposts. Companion plants for this moisture-loving willow are similar to those for pussy willow: Ninebark (*Physocarpus opulifolius*), pagoda dogwood (*Cornus alternifolia*), red osier dogwood (*Cornus stolonifera*), and serviceberry (*Amelanchier canadensis*).

Prairie Willow (*Salix humilis*)

Deciduous Shrub
Size: Height 6–12 feet, width 12–20 feet
Color and Season: Multicolored catkins in early spring; yellow foliage in autumn
Light Requirement: Sun
Soil: Moist to medium dry; loamy, gravelly, sandy
Native Habitat: Prairies, savannas
Native to: All lower 48 states east of Mississippi River plus AR IA MN MO ND NE KS LA OK TX; rare in SD
USDA Hardiness Zones: 3–8

Most often found in savannas, prairie willow is the only native willow shrub found in drier habitats, especially in sandy or rocky soils. Its fuzzy, colorful male catkins have stamens in orange, yellow, purple, and brown and bloom before the leaves emerge. The female catkins are greenish. The spring azure butterfly is often seen nectaring on the flowers. Prairie willow is very fire tolerant and spreads more readily after prescribed burns. As with all willows, the male and female flowers (in catkins) are found on different plants, so it's best

The pine cone gall on this prairie willow has been formed by a small gnat, one of the many insects hosted by willows.
PHOTO BY MARIETTE NOWAK

The common elderberry is gorgeous in blossom with dinner-sized disks of bright white blossoms.
PHOTO BY MARIETTE NOWAK

to plant three to five individuals to ensure pollination and seeds. This species is an excellent selection for drier sites. Companion plants found to grow with this species include lead plant (*Amorpha canescens*), New Jersey tea (*Ceanthus americanus*), and little bluestem grass (*Schzachyrium scoparium*).

ELDERBERRY (*Sambucus* species)

Adoxas or Moschatel Family, Adoxaceae (formerly the Honeysuckle Family, Caprifoliaceae)

Elderberries are medium-sized shrubs with upright spreading branches that arch with age. Some find them graceful, while others consider them leggy and wild in appearance. But everyone marvels at their beauty in flower—tiny white blossoms set in the dinner-plate-sized disks in the common elderberry (*Sambucus canadensis* subspecies *pubens*) or arranged in pyramid-shaped clusters in red elderberry (*S. racemosa* subspecies *pubens*). The fruits that ripen later are gorgeous and bountiful as well, adorning the shrubs with rich purple-black berries and gleaming crimson berries, respectively. And both species are invaluable for birds and other wildlife.

The genus name *Sambucus* refers to an ancient Greek harp called the sambuca. The pith within elderberry stems can be easily removed from the twigs to make whistles and flutes. Indians and early pioneers also used the hollow stems as spiles to tap maple trees for syrup.

Value to Birds and Other Wildlife

The fruits of elderberries are a favorite food of more than twenty species of birds, including the American Robin, Blue Jay, Brown Thrasher, Cedar Waxwing, Eastern Towhee, *European Starling*, Gray Catbird, Mourning Dove, Northern Cardinal, Northern Mockingbird, Pileated Woodpecker, Red-bellied Woodpecker, Red-headed Woodpecker, Ring-necked Pheasant, Rose-breasted Grosbeak, Ruffed Grouse, Song Sparrow, Swainson's Thrush, Veery, White-throated Sparrow, Wood Thrush, and Yellow-breasted Chat. Another thirty-plus species visit to snack on them, including the Alder Flycatcher, American Goldfinch, Bicknell's Thrush, Canada Jay, Chipping Sparrow, Common Grackle, Eastern Bluebird, Eastern Kingbird, Eastern Phoebe, Gray-cheeked Thrush, Great Crested Flycatcher, Greater Prairie Chicken, Hermit Thrush, House Finch, Indigo Bunting, Northern Bobwhite, Northern Flicker, Red-breasted Nuthatch, Red-eyed Vireo, Ruby-crowned Kinglet, Rusty Blackbird, Scarlet Tanager, Swamp Sparrow, Tufted Titmouse, Warbling Vireo, White-breasted Nuthatch, White-crowned Sparrow, White-eyed Vireo, Wild Turkey, Yellow-bellied Sapsucker, and Yellow Warbler.

Many birds also seek cover within their tangled stems. Others, known to build their nests in their sheltering foliage, are the Alder Flycatcher, American Goldfinch, Gray Catbird, Northern Mockingbird, and Yellow Warbler.

Many insects, as well, feast on the shrubs. More than thirty species of caterpillars feed on elderberries and provide nutritious morsels for birds seeking protein for their nestlings. The flowers of elderberries attract many other insects mainly with their pollen, although a bit of nectar has also been reported in common elderberry flowers. Bees, beetles, and flies are the primary pollinators as they gather this pollen. Some bees nest within the pith of the branches. Many other insects feed on various parts of the plants. Some are named for the shrub they love to munch, including the elder aphid, elder shoot borer moth caterpillar, elderberry longhorn beetle larva, elderberry bud midge, elder flower midge, elder stem midge, and elder thrip. All these insects provide a varied diet for birds.

Mammals that are fond of the berries include fox, mice, and squirrels. Because the foliage contains toxic compounds such as cyanide, most mammals avoid it, although deer will browse occasionally on the shrubs. Rabbits chew on the bark as an emergency food in winter.

How to Grow

Common elderberry is the more adaptable of the two species and, although preferring moist soils, grows well in soils of medium to medium-dry moisture, as long as it gets a few hours of sun. Red elderberry is best grown in more northerly locations with cooler climates and is a more shade-tolerant shrub. Both species have rather wild, unkempt appearances but can be heavily pruned in winter, cutting back nearly to the ground, for sturdier branching and better structure.

Propagation

Cuttings and seeds.

Companion Plants

Three herbaceous species—feathery false Solomon's seal (*Smilacina racemosa*), spring beauty (*Claytonia virginica*), and toothwort (*Dentaria laciniata*)—grow with these shrubs in moist open woodlands in the wild and would be beautiful placed beneath them.

A Closer Look

Thrushes are particularly fond of the fruit of elderberries. These fruit lovers include the Swainson's Thrush and Wood Thrush as well as the American Robin

Elderberry fruits are favorites of Swainson's Thrushes and other thrushes.
PHOTO BY JACK BARTHOLMAI

and Veery, which are also in the thrush family. All four are among the major fruit-eating birds in the eastern United States. About 40 percent of their diets consist of berries. Animal matter, such as insects, caterpillars, grubs, and worms (mostly fed on by robins), comprise the other 60 percent of their diets. Elderberries provide an ideal combination of both insects and berries for these birds. In spring and during the breeding season, the birds specialize in insects, usually feeding on the ground. Watch for them foraging in the leaf litter below your elderberries and other shrubs and trees in search of invertebrates. Later in the year, you'll find them up in the foliage foraging for the fruits of elderberry and other shrubs or on the ground eating fallen fruits to fatten up before their fall migrations.

In the summer, be on the lookout for one of our most beautiful native beetles, the elderberry longhorn beetle (*Desmocerus palliatus*). Up to one inch in size, the adult is shimmering deep blue with a bright yellow mid-band (actually its horny forewings). It feeds on the leaves and the pollen of elderberry flowers. The beetle's startling colors warn predators that it is poisonous. The female beetle lays eggs near the base of elderberry shrubs. When they hatch, the larvae bore into the branches of elderberry shrubs and feed on the nutritious wood, roots, and pith of the stems. The elderberries defend themselves with noxious chemicals found in all parts of the plant, but these larvae appear to be immune and retain the toxins into adulthood. Enjoy this beauty, but don't eat it.

Common Elderberry (*Sambucus nigra* subspecies *canadensis*)

Deciduous Shrub
Size: Height 5–10 feet, width 3–8 feet
Color and Season: White flowers in late spring to midsummer; yellow-green foliage in autumn
Light Requirement: Sun, partial shade
Soil: Moist to wet; fertile loam
Native Habitat: Woodlands, marshes, moist prairies
Native to: All lower 48 states (the subspecies *Sambucus nigra* subsp. *canadensis* is native to eastern North America)
USDA Hardiness Zones: 3–9

The fragrant blossoms of common elderberry are arranged in showy flat panicles.

PHOTO BY MARIETTE NOWAK

Common elderberry has long been given the Latin name *Sambucus canadensis*, which many native plant nurseries still use. But recently the shrub has been demoted to a subspecies of black elderberry (*Sambucus nigra*) and given the Latin name *Sambucus nigra* subsp. *canadensis*. The flowers of common elderberry are arranged in flat panicles and are quite fragrant. It blooms in late spring to midsummer for three to four weeks. The berries are deep purple to black and, although inedible raw, are cooked to make tasty jelly, pies, and wine. Common elderberry is fast growing and easy to maintain once established. Prune it regularly to keep it looking neat. It is an excellent choice for naturalizing and screening.

Red Elderberry (*Sambucus racemosa* subspecies *pubens*)

Deciduous Shrub
Size: Height 4–10 feet, width 3–8 feet
Color and Season: White flowers in spring; yellow-green foliage in autumn
Light Requirement: Sun, partial shade
Soil: Moist; sandy loam, silty loam, rocky material
Native Habitat: Moist woodlands, stream banks
Native to: AK AZ CA CO CT DE IA ID IN KY MA MD ME MI MN MT NC ND NH NJ NM NV NY OH OR PA RI SD UT VA VT WA WV WI WY; native but rare in GA IL MO TN; extirpated in DE RI
USDA Hardiness Zones: 3–7

Red elderberry is often referred to as *Sambucus pubens* or *Sambucus racemosa* and these names are still commonly used at native plant nurseries. But most taxonomic authorities have now combined these names for the eastern subspecies and refer to it as *S. racemosa* subsp. *pubens*. The shrub blossoms in mid to late spring with a blooming time of three weeks. Its flowers are arranged in pyramid-shaped panicles and have an unpleasant smell unlike the sweet smell like the common elderberry. The berries have a gorgeous scarlet color and although inedible, even toxic for humans, are readily eaten by birds and mammals. The shrub is more common in cooler, northerly climates than is common elderberry. Fast-growing and tolerant of shade, red elderberry is especially attractive in woodland gardens and can be used for mass plantings and erosion control.

BLUEBERRY (*Vaccinium* species)

Heath Family, Ericaceae

Blueberry shrubs may be deciduous or evergreen. All are lovely in flower and leaf, with delicious super-healthy fruits and interesting winter form. The flowers of blueberries are small urn-shaped bells from which bees hang upside down to pollinate. The berries are deep blue to purple, and the native species usually have smaller and sweeter berries than cultivars. The leaves of different species are variable in size and summer shades of green. In autumn, the foliage

The berries of this early low blueberry are smaller but sweeter than blueberry cultivars.
PHOTO BY PETER M. DZIUK, www.MinnesotaWildflowers.info

of deciduous species is usually spectacular with a palette of flaming reds and oranges, while evergreen species develop more subtle shades of maroon.

Blueberry shrubs are relatively easy to identify but often very confusing to nail down to species. There are many species with lots of variations within a species, as well as multiple natural-occurring hybrids. Naturalist George Ellison calls the confusion "the blueberry identification blues" and, like native plant expert and author William Cullina, refers to two main groups of species: low-bush blueberries, which grow to knee-high, and high-bush blueberries, which are medium-height shrubs or even small trees. Both agree that the low-growing blueberries are especially frustrating. To keep things simple, I'll feature one blueberry species from each group—the early low blueberry (*Vaccinium angustifolium*) and highbush blueberry (*V. corymbosum*), both among the most common species.

Value to Birds and Other Wildlife

Birds love blueberries as much as we do. Hummingbirds nectar on the flowers. The berries are a preferred food for American Robin, Blue Jay, Brown Thrasher, Cedar Waxwing, Eastern Bluebird, Eastern Kingbird, *European Starling*, Gray Catbird, Gray-cheeked Thrush, Eastern Towhee, Hermit Thrush, Northern Flicker, Orchard Oriole, Red-bellied Woodpecker, Red-headed Woodpecker, Ruffed Grouse, Scarlet Tanager, Spruce Grouse, White-throated Sparrow, Veery, Wild Turkey, Wood Thrush, and Yellow-breasted Chat. Grouse eat the leaves as well as the berries.

The rare Kirtland's Warbler often nests beneath jack pines, in the dense cover of blueberry thickets.
PHOTO BY JEREMY MEYER, jeremymeyer.photography

Other birds also known to feed on the berries include American Tree Sparrow, Baltimore Oriole, Black-capped Chickadee, Bobwhite Quail, Eastern Phoebe, Fish Crow, Great Crested Flycatcher, Greater Prairie Chicken, Herring Gull, Mourning Dove, Northern Cardinal, Pine Grosbeak, Ring-necked Pheasant, Sandhill Crane, Savannah Sparrow, and Tufted Titmouse.

Many birds also find protective cover in the shrubs. Among those known to do so are American Tree Sparrow, Gray Catbird, Gray-cheeked Thrush, Eastern Towhee, Hermit Thrush, Mourning Doves, Northern Bobwhite, Ring-necked Pheasant, Ruffed Grouse, Spruce Grouse, Veery, White-throated Sparrow, Wild Turkey, and Wood Thrush. This cover is also important for nesting birds, especially catbirds, for which it is a favorite nest site, and some warblers (see later in this section). An amazing 297 butterfly and moth caterpillar species have been documented feeding on blueberries! They, in turn, provide juicy protein-packed "burritos" for nestlings. The butterfly caterpillars are those of the brown and Henry's elfins, green comma, pink-edged sulphur, spring azure, and striped hairstreak. Among the many moth caterpillars are geometers, the familiar inchworms that loop their backs as they move, slender clearwing moths that look like tiny hummingbirds as they nectar on flowers, and blueberry cinder moths named after the plant they devour.

Also of great importance are the many bees that visit the flowers for pollen and nectar, pollinating them while doing so. Both honeybees, which were originally imported from Europe, and native bumblebees pollinate blueberries. But bumblebees have been found to be the champion pollinators in a study led by University of Vermont scientists. The bumblebees increased both the size and amount of blueberries produced. This was attributed to their buzz pollination in which they vibrate the flowers to release showers of pollen. Honeybees can't do this and instead pry out the pollen. The study highlights the need to protect and provide habitat for wild bees, which are in alarming decline throughout the country. Because of the decline in native bees, honeybee hives are often brought in to pollinate blueberries.

Quite a number of other insects that feed on other parts of blueberry plants are named after the shrub: blueberry aphid, blueberry bud gall midge, blueberry fruit fly, blueberry fruit midge, blueberry gall midge, blueberry leaf gall midge, blueberry mealybug, and blueberry stem gall wasp. There are, of course, additional insects found on blueberries, including beetles, grasshoppers, leafhoppers, spittlebugs, and weevils. With all these voracious critters, it's good to know birds are helping to keep them in check.

Terrestrial turtles—eastern box turtles and wood turtles—munch on the berries, as do most mammals. Bears are said to be especially fond of blueberries, but chipmunks, fox, mice, skunks, and squirrels also eat the fruits. Deer and rabbits browse on the leaves and twigs.

How to Grow

Most blueberries need strongly acidic soils with a pH between four and five. They spread on rhizomes to form colonies, making them an excellent choice for a ground cover, and their shallow roots make them easy to divide and transplant. Most species do best in a sunny location for good berry production, but some grow in dry, shady locations. Because they are somewhat self-infertile, it's best to install several plants that can cross-pollinate one another. Blueberries are fire-tolerant plants. Growers have traditionally burned their blueberry fields frequently to eliminate competing plants and to fertilize the soil. Blueberries flower on two-year-old stems, however, so burns, including prescribed burns for restoration purposes, will set the flowering and fruiting back a year.

Propagation

Seeds and cuttings.

Companion Plants

Some of the plants that are found growing with blueberry shrubs in the wild are woodland sunflower (*Helianthus divaricatus*), savanna sunflower (*Helianthus strumosus*), and wild geranium (*Geranium maculatum*).

A Closer Look

Several rarely seen warblers, the elusive Connecticut Warbler and the rare Kirtland's Warbler, often nest on or near the ground in blueberry thickets and other low vegetation. Although both primarily forage on insects, they also eat blueberries and other fruits. You may catch a glimpse of the secretive Connecticut

The secretive Connecticut Warbler skulks around under blueberry thickets, gleaning insects and eating blueberries.
PHOTO BY JEREMY MEYER, jeremymeyer.photography

Warbler in spring on its way from South America to the north woods of Michigan, Minnesota, Wisconsin, and Canada. One spring day a few years ago, I had the rare pleasure of having one stop briefly in my yard, where it foraged in the leaf litter beneath oaks. If you live in or visit northern woods, you may be lucky to discover one skulking about in an understory of blueberry or other dense cover beneath jack pines. Its loud *chip-chup-ee* song can be a clue to its presence. Sadly, lack of habitat both here and in South America has caused Connecticut Warbler numbers to plummet throughout its range.

Kirtland's Warblers are rarely seen during their migration between their wintering grounds in the Bahamas and other Caribbean islands and their breeding grounds in Michigan, Wisconsin, and Ontario. Like avian prima donnas, these warblers breed only in young jack pine forests, placing their nests deep within the dense cover of understory plants like blueberries. The best way to see this beautiful warbler is to join a guided tour, as my husband and I have done.

Also not to miss is the fascinating buzz pollination of blueberry flowers by bumblebees. The bees clasp the pollen-bearing anthers with their legs or mouthparts and then vibrate at a high frequency to shake out the pollen. This releases a large amount of pollen from one flower that would ordinarily take visits to numerous flowers to gather. Other food crops that bumblebees buzz-pollinate include cranberries, tomatoes, eggplants, and peppers. Partridge pea (*Chamaecrista fasciculata*) and wild senna (*Senna hebecarpa*) are native wildflowers that are buzz-pollinated.

Early Low Blueberry (*Vaccinium angustifolium*)

Deciduous shrub
Size: Height 8–24 inches, width 1–3 feet
Color and Season: White tinged with pink flowers in spring: fire-red foliage in autumn
Light Requirement: Sun, partial shade
Soil: Moist to dry; acidic sandy, well drained light soils or rocky soils
Native Habitat: Shrubby savannas, barrens, openings in pine forests, rocky cliffs
Native to: CT IL IN MA MD ME MI MN NH NJ NY OH PA RI VA VT WI WV; native but rare in IA NC; extirpated in DE
USDA Hardiness Zones: 2–8

Early low blueberry is the beloved blueberry of Maine, but it is also found throughout most midwestern states as well as eastern states and is one of the prime sources for commercial blueberries. Early low blueberry is a good choice for a low ground cover with brilliant red fall color. The species epithet *angustifolium* is a combination of the Latin words *angustum* (narrow) and *folium* (leaf), referring to the narrow leaves. This species is very similar to hillside or late low blueberry (*V. pallidum*) with confusing intermediate forms and hybrids of the two. The late low blueberry typically has larger oval leaves that are paler beneath, as suggested by *pallidum* in its Latin name. There are other nearly identical species in the west and southeast.

Highbush Blueberry (*Vaccinium corymbosum*)

Deciduous Shrub
Size: Height 3–10 feet, width 3–10 feet
Color and Season: Pink-tinged white flowers in spring; bright red foliage in autumn
Light Requirement: Sun, partial shade
Soil: Wet to moist, acidic soil that is sandy or peaty
Native Habitat: Bogs, swamps, wetlands
Native to: AL AR CT DE FL GA IN KY LA MA MD ME MI MO MS NC NH NJ NY OH OR PA RI SC TN TX VA VT WA WV; native but rare in IL OK WI
USDA Hardiness Zones: 5–8

As its name suggests, this species is tall and upright. Although in the wild it grows in bogs and acidic wetlands, it does well in gardens with moist, acidic soils and is often grown for both its ornamental value and delicious berries. The berries are blue black with a whitish bloom, about a third inch in diameter, and have a sweet to sweet-tart taste. Highbush blueberry is an excellent choice for hedges or shrub borders.

VIBURNUM (*Viburnum* species)

Adoxas or Moschatel Family, Adoxaceae (formerly the Honeysuckle Family, Caprifoliaceae)

Viburnums are a large and diverse group of shrubs with clusters of showy flowers, colorful fruits, and attractive fall color. Because of their beauty through the seasons, they are some of the most popular shrubs for landscaping. They also provide abundant food for birds and other wildlife.

Various viburnum species are suitable for a variety of different growing conditions. Some do well in shade; others prefer sun. Some like wet to moist soils; most prefer moderately moist soils but are tolerant of dry soils once established.

Viburnums, like this black haw, host more than one hundred caterpillar species—a critical food for nestlings.
PHOTO BY MARIETTE NOWAK

Some need richer soils, while others grow well in nutrient-poor soil. Many spread and form thickets; others are less aggressive and can be grown singly as specimen plants. They also have many different leaf shapes—lobed maple leaf mimics, lance shaped to round, toothed to entire (nontoothed). But one thing they all have in common are leaves that are opposite, not alternate, on the branches.

Despite their differences, all viburnums are generally easygoing shrubs for the garden or for wildlife restoration. In the past, I have championed many species of viburnum. But unfortunately, an invader from Europe, the viburnum leaf beetle (*Lymantria dispar*), has decimated some species. Consequently, I will feature only the two native viburnums that are the most resistant to the beetle menace: nannyberry (*Viburnum lentago*) and black haw viburnum (*V. prunifolium*). I have both in my yard, and they are healthy and free of the beetle. I also have several of the more susceptible species, including American cranberry bush (*V. trilobum*), European cranberry bush (*V. opulus*), and downy arrowwood (*V. rafinesquianum*). Of these, only several European cranberry bush shrubs (part of the original landscaping, not planted by me) have some beetle damage. The viburnum beetle is relatively new to the Midwest and is "wreaking havoc in some areas and still expanding into others," according to PJ Liesch, University of Wisconsin–Madison entomologist. He advises against planting arrowwood viburnum (*V. dentatum*) or American cranberry bush (*V. opulus* var. *americanum*; also named *V. trilobum*), which are strongly preferred by the beetle.

The good news is that there are fewer outbreaks occurring in Canada and New York state, where the beetle first gained ground, according to Brian Eshenaur, Cornell University entomologist. He believes that "there could be hope for the future, as it seems, a few years after viburnum leaf beetles move into an area, a decline and leveling off in the population occurs." He suspects that beneficial insects may be helping to keep the beetles in check. For the time being, however, it's best to stick to the least susceptible native viburnums, the beautiful nannyberries and black haws.

Value to Birds and Other Wildlife

Viburnum berries have a lower fat content than those of dogwoods. As a result, they are less prone to rot and last longer. The sequence of these two widespread groups of shrubs is very advantageous to birds. The early migrants tend to quickly strip the dogwoods of berries, while viburnum berries provide for the later migrants and resident birds, helping to extend the berry season into early winter and sometimes even into spring. Many birds feast on viburnum berries, including the American Robin, Bicknell's Thrush, Blue-headed Vireo, Brown Thrasher, Canada Jay, Cedar Waxwing, Eastern Bluebird, *European Starling*, Gray Catbird, Gray-cheeked Thrush, Great Crested Flycatcher, Hermit Thrush, Northern Bobwhite, Northern Cardinal, Northern Flicker, Northern Mockingbird, Pileated Woodpecker, Pine Grosbeak, Purple Finch, Red-eyed Vireo, Ring-necked Pheasant, Rose-breasted Grosbeak, Ruffed Grouse, Spruce Grouse, Swainson's Thrush, White-throated Sparrow, Wild Turkey, and Wood Thrush.

The dense shrubs also provide cover for birds. Those known to shelter in viburnums include Bicknell's Thrush, Gray Catbird, Hermit Thrush, Northern Bobwhite, Ruffed Grouse, Ring-necked Pheasant, Spruce Grouse, Swainson's Thrush, Willow Flycatcher, Wild Turkey, and Wood Thrush. Birds that nest in the shrubs include the Gray Catbird, Indigo Bunting, Prairie Warbler, White-eyed Vireo, Willow Flycatcher, and Wood Thrush.

Viburnum pollen and nectar attract many small bees and flies. Butterflies, skippers, hummingbird moths, and ants visit for nectar. More than one hundred caterpillar species feed on various parts of viburnums and provide food for birds. Among them are those of the lovely spring azure, which feed on the flowers and buds. The hummingbird clearwing moth and several sphinx moth caterpillars also feed on viburnums.

Mammals likewise find nourishment in viburnums. Chipmunks, fox, mice, and squirrels eat the berries. Rabbits gnaw on the bark. Beavers eat the wood. And deer browse on the twigs and leaves. The berries are edible and tasty for humans too.

How to Grow

Both nannyberry and black haw have upright forms with rounded crowns. They are usually multistemmed shrubs but can be pruned to form small single-trunk trees by removing stems and suckers at the base. Use as specimen shrubs or allow them to form colonies for privacy screens or shrub borders. Even though their flowers have both male and female parts, viburnums are self-infertile. Two or more genetically different shrubs are needed for cross-pollination and fruiting. Purchase from a nursery that produces plants from seed or buy from different nurseries that source from different growers.

Propagation

Seeds usually require a two-year germination period.

Companion Plants

Herbaceous species can provide a ground cover at the base of the shrubs. Some known to grow with viburnums in the wild are easy-to-grow plants like common oak sedge (*Carex pensylvanica*), wild geranium (*Geranium maculatum*), and Virginia waterleaf (*Hydrophyllum virginianum*).

A Closer Look

After the migrants have left for the winter, watch for wintering birds like Wild Turkeys, Northern Cardinals, and Cedar Waxwings seeking out the last of the viburnum's berries. It's also a good time to check among the dense branches for the nests of Indigo Buntings, Gray Catbirds, and other shrubland birds.

Since the viburnum leaf beetle invasion, it's also wise to give your viburnums a close look throughout the year, especially if you have susceptible species. Both beetle larvae and adults feed on the leaves. In spring, the larvae hatch and are one millimeter long, greenish yellow to off white, and lack spots. Later stage larvae are yellowish brown with black spots along their backs and grow to about a half inch. When small, they are harder to spot and often feed on the underside of leaves, usually in a group. When larger, the larvae can often be spotted on the tops of leaves, but they drop quickly when disturbed. From

Indigo Buntings often nest within the dense branches of viburnums.

PHOTO BY JEREMY MEYER, jeremymeyer.photography

spring to early summer, the larvae feed on the leaves between the veins, creating many small holes. Then they crawl down to the ground to pupate. About ten days later, the brown adults, one-quarter inch in size, emerge and continue to feed on the leaves, biting holes in them. In late fall, the females lay eggs on the twigs. The best way to combat this beetle is to destroy the egg-laying sites, as I've done. Look for them on the tips of twigs. They are tiny football-shaped bumps the size of a sesame seed, usually on the underside of the twigs and laid in rows up to several inches long. In late fall or winter, prune off these twigs and trash them (don't compost). Avoid spraying because you may kill beneficial insects as well as the beetles. Ladybugs, soldier bugs, green lacewings, stink bugs, and assassin bugs all eat viburnum beetle larvae and some eat the adults too. Support these beneficial insects by planting a diversity of native plants. For severe infestations, horticultural soaps have been effective in killing larvae but only when the larvae are very small.

Nannyberry (*Viburnum lentago*)

Deciduous Shrub
Size: Height 8–15 feet, width 6–12 feet
Color and Season: White flowers in spring; burgundy foliage in autumn
Light Requirement: Sun, partial sun
Soil: Moist to dry; well-drained soils, tolerant of clay
Native Habitat: Low moist woodlands, prairies, marshes
Native to: CT IA IL IN MA ME MI MN ND NH NJ NY OH PA RI SD VT WI WY; native but rare in DE KY MD MO MT NC NE WV; extirpated in VA
USDA Hardiness Zones: 2–8

One of the largest viburnums, nannyberry is a beauty throughout the year with a profusion of spring flowers, dark green summer foliage, blue-black berries, and burgundy fall color. One of the distinct features of this versatile shrub—a good identification clue—is its long, beige terminal bud. It does best in moist soils and at least a half day of sun. This shrub is called nannyberry or sheepberry because some think it smells like a sheep or goat (I've never noticed that it does). Its delicious berries, on the other hand, have inspired a more complimentary name, sweet viburnum. The berries taste like raisins, prunes, dates,

The edible fruits of nannyberries change color gradually from greenish yellow to bluish black when fully ripe.

PHOTO BY MARGARET KNIGHT

bananas, or prunes, depending on one's palate, and are sweetest after a frost. Nannyberry spreads readily by roots, a trait that can be advantageous for stabilizing shorelines, slopes, and stream banks as well as creating hedges and screens. I planted mine under light shade, where they've expanded to create a dense, shrubby haven for birds. Nannyberry can also be used as a handsome specimen shrub or small tree by pruning out suckers and lower branches.

Black Haw Viburnum (*Viburnum prunifolium*)

Deciduous Shrub or Small Tree
Size: Height 8–15 feet, width 6–15 feet
Color and Season: White flowers in spring; reddish-orange foliage in autumn
Light Requirement: Sun, partial sun

After pollination, the beautiful flowers of viburnums are followed by berries beloved by birds.
PHOTO BY MARIETTE NOWAK

Soil: Moist to dry; tolerant of most soils once established, drought and alkaline tolerant
Native Habitat: Thickets, woodlands, savannas
Native to: AL AR DE GA IL IN KS KY LA MA MD MO MS MT NC NJ NY OH OK SC TN TX VA WV; native but rare in CT IA MI WI
USDA Hardiness Zones: 3–9

Black haw viburnum is gorgeous all seasons. Its creamy white blossoms are abundant and showy in late spring and have a sweet fragrance when fresh. In late summer and fall, it displays attractive blue-black berries and bright fall color. Its oval leaves resemble those of wild plums, which explains the species name *prunifolium*. Black haw is much like nannyberry in size, flowers, and berries. Both are very adaptable and easy to grow. But black haw viburnum can handle drought, poor soils, and light shade better, while nannyberry does best in moist, richer soils and sunnier sites. Black haw also has more sturdy and crooked branching, giving it a hawthorn-like appearance, especially in winter. Black haws can be grown as specimens or grouped in shrub borders, hedges, or screens. I planted one as a specimen shrub beside my bird-feeding area and enjoy seeing the birds sunning and perching among the branches while they wait their turn for feeder goodies. A few years ago, in order to encourage fruiting, I added several more black haws and now have a small copse of them.

Yellow Warbler.

PHOTO BY JEREMY MEYER, jeremymeyer.photography

8 Top 10 Plants for Warblers and Other Woodland Birds

Multitudes of birds court, feed, roost, nest, and raise their young in trees, whether in single specimen trees, small leafy groves, woodlots, or expansive forests. Some birds prefer younger trees, while others seek out old growth trees. Warblers, for example, are generally birds of woodlands and forests but vary widely in the habitats they prefer. Many breed in conifers. Others nest in deciduous forests or mixed conifer/deciduous forests. And birds sometimes show a strong preference for particular tree species, as you'll see with some of the trees highlighted in this chapter.

The tree species featured here are among those that offer the best habitat and foods for various species of woodland birds. There are trees such as birch, maple, and poplar that offer seeds for birds. Others provide nuts, including hickory and oak. Cherries and crabapples produce berries for fruit-eating birds. Whatever the species, it is critical to avoid planting trees that are seedless cultivars—sterile trees that offer none of these foods for birds. Unfortunately, there are seedless maples and aspens on the market, as well as crabapple cultivars that don't produce berries. As always, buyer beware.

Trees have been described as "meadows in the sky" because of their masses of flowers in spring. In that season, trees attract myriads of insects, as well as birds, that visit to feast on the bounty. Juicy, nutritious caterpillars are especially treasured finds for most birds. The trees featured here are among those that host the greatest number of butterfly and moth caterpillars, based on the research of Douglas Tallamy, University of Delaware entomologist. Caterpillars are essential food for the nestlings and favored foods for adult insectivorous

Some Breeding Birds of the Woodlands of the Midwest and Eastern United States

This Baltimore Oriole is building its hanging nest in a box elder tree.

PHOTO BY JACK BARTHOLMAI

Chickadees: Black-capped and Carolina

Chuck-will's-widow

Brown Creeper

Red Crossbill

Cuckoos: Yellow-billed and Black-billed

Purple Finch

Flycatchers: Acadian, Great Crested, Least, Olive-sided, and Yellow-bellied

Blue-gray Gnatcatcher

Grosbeaks: Evening and Rose-breasted

Ruffed Grouse

Hawks: Broad-winged, Cooper's, Red-shouldered, and Sharp-shinned

Ruby-throated Hummingbird

Canada Jay

Dark-eyed Junco

Eastern Kingbird

Kinglets: Golden-Crowned and Ruby-Crowned

Nuthatches: Red-breasted and White-breasted

Orioles: Baltimore and Orchard

Ovenbird

Owls: Barred, Screech, and Great Horned

Northern Parula

American Redstart

Yellow-bellied Sapsucker

Pine Siskin
Tanagers: Scarlet and Summer
Thrushes: Hermit, Swainson's, and Wood
Tufted Titmouse
Wild Turkey
Veery
Vireos: Blue-headed, Red-eyed, Yellow-throated, and Warbling
Warblers: Black-and-white, Blackburnian, Black-throated Blue, Black-throated Green, Canada, Cape May, Cerulean, Hooded, Kentucky, Magnolia, Pine, Prothonotary, Tennessee, Worm-eating, Yellow-rumped, and Yellow-throated
Waterthrushes: Louisiana and Northern
Whip-poor-will
Woodpeckers: Downy, Hairy, Pileated, and Red-bellied
Eastern Wood-Pewee
Winter Wren

birds. Trees continue to host a great variety of caterpillars and other insects throughout the year, even in winter when birds can find insects hidden in crevices in the bark of trees, drilled into the wood, curled in dead leaves, or nestled in leaf litter.

A soft landing beneath trees is important for insects. Many lay their eggs on the foliage, branches, or bark of trees. After hatching, the larvae feed their fill high in the trees. Many insects then drop down to the ground and overwinter as pupae in the leaf litter and soil. Adding native plants beneath the trees will create a soft landing for these insects where they can overwinter in the plant debris and fallen leaves. Suggestions for the best herbaceous plants to grow beneath the trees are provided for each genus of trees in this chapter.

MAPLE (*Acer* species)

Soapberry Family, Sapindaceae

Maples are among the most recognized and beloved species in the landscape, ranging from shrubs and small understory trees to majestic forest canopy trees. They are easily identified by their palmately lobed leaves (an exception is box elder, *Acer negundo*). All have opposite leaves and winged seedpods called samaras.

Maples may have either male flowers or female flowers on the same or different trees. The male flowers are the only ones that have nectar and pollen to attract bees and other pollinators. The female flowers offer no floral rewards to insects and are seldom visited by bees and other pollinators, so cross-pollination by insects doesn't occur. Maple trees, instead, rely on the wind to blow their pollen onto sticky receptive female flowers to fertilize them and start seed production.

As the fertilized female flowers fade, they begin forming seeds and package them within samaras, paired flattened papery pods, each containing a seed. The samaras, often called helicopters or maple keys, twirl away in the wind, sending them some distance from the heavy shade of the parent trees. During World War II, the U.S. Army designed an airdrop supply carrier based on the

The sugar maple has the most magnificent fall color of any tree in North America.
PHOTO BY MARIETTE NOWAK

samara. Sadly, many maples in our landscapes are all-male cultivars developed to avoid "messy" samara litter, but they may also contribute to higher allergy-inducing pollen counts. Even worse, in our obsession with litter-free landscapes, children no longer delight in falling helicopters, and birds and squirrels are no longer nourished by their seeds.

Four of the most common maple trees are featured here: box elder or ash-leaved maple (*A. negundo*), red maple (*A. rubrum*), silver maple (*A. saccharinum*), and sugar maple (*A. saccharum*).

Two of these species, sugar maple and red maple, have a high content of sugar in their sap and are important in maple syrup production. Sugar maple averages 2.0 to 2.5 percent sap sugar content, while red maple has about half that at 1.5 to 2.0 percent sap sugar. In the strong winds of late winter, maple limbs are sometimes sheared off. Sticky, sweet-tasting globs and icicles soon form at the broken stubs as the sap begins to rise in spring. American Indians soon learned to harvest and boil this sap into syrup, which they added as a sweetener to many foods, including meats. When traveling, they carried maple sugar cakes wrapped in bark for an easily portable food, much like today's granola bars. While not touted as a health food, pure maple syrup contains more nutrients, including antioxidants and bone-building manganese, magnesium, and phosphate, than sugars.

Value to Birds and Other Wildlife

Maples are invaluable for birds, providing for nearly all their needs—seeds, insects, cover, and nest sites. First in line at the maple table are early migrating Yellow-bellied Sapsuckers, who drill holes in bands around sugar maples and feast on their upwelling sap with brush-tipped tongues. Ruby-throated Hummingbirds soon visit to sip the sugary sap, a vital food when few nectar-rich flowers are in blossom. The sap also attracts butterflies that hibernate as adults and emerge early in spring, including the eastern comma, mourning cloak, question mark, and red-spotted purple, all likely famished for food after their winter's sleep.

Maple seeds are available throughout most of the year, as different species ripen their samaras, in sequence. Among those maples featured here, silver maples are the first to set seed, their samaras ripening as early as April and

Yellow-bellied Sapsuckers drill holes in maples to feast on the upwelling sap.
PHOTO BY JACK BARTHOLMAI

May. The next samaras to mature are those of the red maple, then box elder, and last the sugar maple samaras, which ripen from late summer to September. The box elder samaras often persist into winter, providing survival food for wintering birds. I've even seen very early spring migrants feeding on the last remaining box elder samaras when little else was available.

Grosbeaks, especially the beautiful Pine, Rose-breasted, and Evening Grosbeaks, as well as Purple Finches flock to maples to gobble up their seeds, one of their favorite foods. Other birds that join in the feast at times are the American Goldfinch, American Robin, Carolina Chickadee, Northern Bobwhite, Northern Cardinal, Pine Siskin, Ring-necked Pheasant, Ruffed Grouse, Song Sparrow, and Wild Turkey. Two species, the Evening Grosbeak and Purple Finch, enjoy maple buds as well as seeds. The Red-breasted Nuthatch and White-throated Sparrow are known to indulge only in the buds. Squirrels occasionally eat the seeds, and rabbits munch on seedlings.

Maples are also popular nesting sites for birds such as the American Goldfinch, American Robin, Baltimore Oriole, Evening Grosbeak, Prairie Warbler, Red-eyed Vireo, Rose-breasted Grosbeak, and Yellow Warbler, who nestle in their leafy boughs. The Red-bellied Woodpecker, *European Starling*, Screech-Owl, Pileated Woodpecker, Wood Duck, Northern Flicker, and Tree Swallow nest in cavities in trees. Tree squirrels and bats also use the cavities. The foliage and twigs are browsed by deer, elk, and moose in winter when other food is scarce.

Throughout the growing season, a great variety of insects feed on maples, including the nearly three hundred caterpillar species. An impressive number of them are such regulars that they are named after the trees. Among moths, there's the rosy maple moth, maple twig borer moth, maple bud borer moth, lesser maple spanworm, maple leaf skeletonizer moth, maple bud borer moth, orange-humped mapleworm, maple leaf blotch miner moth, lesser maple leaf-blotch mine moth, maple leaf miner moth, and maple leafcutter moth. Several specialize in box elders, including the box elder leafroller moth, large box elder leafroller moth, and box elder twig borer moth. Many other insects also carry the name of their favorite food plant, including the maple mealybug, maple gouty vein-gall midge. painted maple aphid, cottony maple leaf scale, maple scurfy scale, maple bladder gall mite, maple scurfy scale, maple spindle gall, box elder bud gall midge, box elder bug, box elder aphid, box elder psyllid, and box elder pouch ball mite.

In turn, hungry birds gorge on these critters. The Prothonotary Warbler and the Acadian Flycatcher are the most voracious foragers on insects in mature maples. Other species known to seek out maple insects are the Blue-gray Gnat-catcher, Carolina Chickadee, Cerulean Warbler, Northern Parula Warbler, Scarlet Tanager, Summer Tanager, Yellow-billed Cuckoo, Yellow-throated Vireo, and Yellow-throated Warbler.

How to Grow

Maples are fast growing and generally prefer moist, slightly acidic soil, although they are quite adaptable to a wide range of conditions. Plant them as specimen trees or in small groups or in woodland restorations. They have shallow roots, making them easy to transplant but also causing damage to the surfaces of streets and roads. Sugar and red maples, and to a lesser extent silver maple and box elder, are also very sensitive to salt. The salt kills the mycorrhizal fungi in the soil, which supply maples with the nutrients they need. Because of maples' shallow roots and intolerance to salt, it's best to avoid planting them along streets and roads. Maples are susceptible to some diseases like anthracnose, but these do not usually cause severe problems. Nor are maples greatly harmed by most insects that feed on them, which are usually kept in check by beneficial insects and birds. One, however, the Asian long-horned beetle (*Anoplophora glabripennis*), has killed many maples. Fortunately, it is one of the few invasive species that has been eradicated in many areas (albeit at the cost of millions). The latest data shows infestations remain in four states: Massachusetts, New York, Ohio, and South Carolina.

Propagation

Seeds.

Companion Plants

All maples produce heavy shade, which greatly limits plant growth beneath them. Early in spring, however, the understory of native maple woodlands is often carpeted with a profusion of the blossoms of ephemeral plants. These are herbaceous species that bloom and set seed in a few short weeks before becoming shaded by the expanding maple foliage. The ephemerals then die back and disappear completely without a trace, having stored food underground in corms, tubers, rhizomes, or bulbs for next spring's growth. Common species include false rue anemone (*Isophyrum biternatum*), toothwort (*Dentaria laciniata*), and trout lilies (*Erthronium albidum* and *E. americanum*). Some other common understory species in maple woodlands also bloom in early spring but retain their leaves and ripen their fruit in summer, including bloodroot (*Sanguinaria canadensis*), hepatica (*Hepatica acutiloba*), and trilliums (*Trillium grandiflorum and T. recurvatum*).

A Closer Look

Keep an eye out for rows of small holes on the trunks of maples. These are the sap wells dug by Yellow-bellied Sapsuckers to start the flow of sap on which they feed. Sugar maple trees are a favorite source of sap for yellow-bellies. The birds don't actually suck up the sap, as their name implies; instead, they use their extremely long, bristly, feather-tipped tongues to lap it up, along with any trapped insects. You have a good chance of seeing the sapsuckers since they tend the sap wells, working tirelessly to maintain them by pecking and licking at them so they don't dry up. You may also see Ruby-throated Hummingbirds, which have a close relationship with sapsuckers. The hummingbirds seem to time their migration to that of the sapsuckers and survive on the sap tapped by the woodpeckers when nectar-rich flowers are not yet available. The hummers often nest near sap wells for easy access to them throughout the breeding season. The sap wells are also important foraging sites for many insects, including moths, butterflies, beetles, and flies comprising twenty families in all. Research has shown that the presence of sapsuckers has a strong influence on the size, composition, and distribution of local insect communities as well as those of bird communities.

Like those of so many trees, maple flowers are easily overlooked. Be sure to watch for the varied colors and forms of their inflorescences in spring. Even from a distance, spring woodlands are often aglow with pale yellows and reds radiating from blossoming maples. Up close, look for the reddish blooming tassels at the tips of the twigs on red and silver maples. In contrast, boxwood has graceful flowering chandeliers, soon followed by those of sugar maple. But don't expect any showy petals. Only red maple has petals, minute as they may be, while the flowers of the other three maples are simply stamens and sticky pistils waiting for wind-blown pollen.

Box Elder or Ash-Leaved Maple (*Acer negundo*)

Deciduous Tree
Size: Height 20–40 feet, width 15–30 feet
Color and Season: Reddish flowers in spring; yellow to yellow-green foliage in autumn
Light Requirement: Sun to partial shade
Soil: Moist to dry; tolerates most soils, including heavy clay, sand, deep loam, or gravelly soils
Native Habitat: Floodplain woodlands, disturbed areas
Native to: All lower 48 states
USDA Hardiness Zones: 2–10

Box elder is a relatively small tree with a short trunk branching into several low, stout branches—perfect for tree-climbing kids. It grows fast and is extremely tolerant of most soils—clay, sand, loam, or gravelly. Unlike other maples, it has pinnately compound leaves with three to seven leaflets similar to an ash tree, hence the alternate name *ash-leaved maple*, and it has smooth branches with a distinct olive-green color. Male flowers and female flowers occur on different trees.

Box elder is a much-maligned tree due to its weedy tendencies and susceptibility to wind damage. Despite these drawbacks, I cherish the several I have in my yard because I admire their flowering "chandeliers" in early spring and the birds and wildlife they support throughout the year. In winter, their asymmetrical branching gives them a handsome, rugged appearance.

The box elder bug (*Boisea trivittata*), which breeds on box elder, deserves mention since it can become a nuisance in some years. In fall as the weather cools, these black bugs with stylish red stripes swarm on the sunny southwestern sides of buildings and often try to escape into our homes through an open door or a cracked screen. This is not a big problem, however, as these harmless bugs can be easily vacuumed or swept up.

Box elder is an excellent choice for stabilizing riverbanks and for difficult sites. With some pruning, it's a good choice for a small yard. And if you have children, it's a great natural substitute for monkey bars and tree forts.

Red Maple or Swamp Maple (*Acer rubrum*)

Deciduous Tree
Size: Height 40–75 feet, width 20–50 feet
Color and Season: Red flowers in early spring; red, orange, yellow foliage in autumn
Light Requirement: Sun, partial shade
Soil: Wet to moist; acidic, tolerant of clay
Native Habitat: Wet to moist woodlands
Native to: AL AR CT DE FL GA IL IN KY LA MA MD ME MI MN MO MS NC NH NY OH OK PA SC TN TX VA VT WI WV
USDA Hardiness Zones: 3–9

Red maple lives up to its name, showing red color every season of the year. It has reddish twigs in winter, red flowers in spring, red samaras, and crimson red fall foliage (although the color is variable). It is an excellent choice for a moderately sized specimen or shade tree and has a wide, rounded crown and smooth gray bark that sets off its red twigs. Although it prefers moist soils, red maple is quite adaptable to average soils. It can grow as much as two feet a year, faster than sugar maple. Three large red maples grow in my neighborhood not far from me, but their branches are never decorated with showy red samaras. The trees are likely seedless cultivars, and their lack of samaras deprives us of one of the beautiful stages in their life cycle and, more importantly, deprives Evening Grosbeaks, a declining species, of one of their favorite foods.

Silver Maple (*Acer saccharinum*)

Deciduous Tree
Size: Height 60–75 feet, width 30–50 feet
Color and Season: Red flowers in early spring; yellow-green to yellow foliage in autumn
Light Requirement: Sun to partial shade
Soil: Average to wet; tolerant of clay, alkaline and acidic soils
Native to: AL AR CT DE FL GA IA IL IN KS KY LA MA MD ME MI MN MO MS NC NE NH NJ NY OH OK PA RI SC SD TN TX VA VT WI WV
USDA Hardiness Zones: 3–9

Silver maple was once a popular street tree due to its fast growth and attractive foliage with silvery undersurfaces that flash in the breeze. But its shallow roots can damage sidewalks, and its very weak wood causes its brittle branches to crash down in storms and damage nearby homes. For wildlife, however, this weakness is a bonanza, creating cavities that can be used as nest sites by owls and Wood Ducks and as dens by raccoons, squirrels, and opossums. Birdwatchers report that their backyard silver maples are magnets for birds. In spring, warblers visit to snap up the clouds of gnats swarming the flowers, and Red-bellied Woodpeckers lap up sap running from fissures in the bark. Throughout the year, birders enjoy seeing cuckoos, Brown Creepers, flycatchers, woodpeckers, and many other species feasting on insects. All agree, it's a stellar tree for birds.

Sugar Maple (*Acer saccharum*)

Deciduous Tree
Size: Height 60–80 feet, width 25–50 feet
Color and Season: Yellow-green flowers in spring; orange, gold, and crimson foliage in autumn
Light Requirement: Sun, partial shade, or full shade
Soil: Moist; rich upland soils, intolerant of soggy soils
Native Habitat: Forests
Native to: AL AR CT DE GA IA IL IN KS KY MA MD ME MI MN MO MS NC ND NE NH NJ OH OK PA RI SD TN VA VT WI WV
USDA Hardiness Zones: 3–8

> The most magnificent display of color in all the kingdom of plants is the autumnal foliage of the trees of North America. Over them all, over the clear light of the Aspens and Mountain Ash, over the leaping flames of Sumac and the hell-fire flickerings of poison ivy, over the war-paint of the many Oaks, rise the colors of one tree—the Sugar Maple—in the shout of a great army. Clearest yellow, richest crimson, tumultuous scarlet, or brilliant orange—the yellow pigments shining through the over-painting of the red. It is like the mighty, marching melody that rides upon the crest of some symphonic weltering sea and, with its crying song, gives meaning to all the calculated dissonance of the orchestra.
>
> —Donald Culross Peattie, *A Natural History of Trees*

One of the most beloved trees in the eastern half of the country, sugar maple is especially renowned for its brilliant fall foliage, as so exuberantly described by Donald Culross Peattie, and for the delicious maple syrup produced from its sap. It is a dominant tree in southern mesic forests, preferring rich, moist soils. One of the hardiest of all maples, it can live as long as four hundred years or more. A stately tree with a dense, rounded form, it is a top choice for a shade or specimen tree. But because it is very sensitive to salts, pollutants, and compacted soils, this species is unsuitable for use as a street tree. Black maple (*A. nigrum*) is very similar to sugar maple. Some experts consider it a separate species, while others lump it with sugar maple, occasionally labeling it *A. saccharum* var.

nigrum. Sugar maple's alternative common name is hard maple, because of the great strength of its timbers. It is one of most prized species for flooring, sports products, and musical instruments, such as violins and guitars. Of particular interest is bird's-eye maple, a rare wood pattern found primarily in sugar maple. The pattern resembles tiny curving eyes of birds, with shapes unique to each particular tree, and is highly valued for its beauty.

BIRCH (*Betula* species)

Birch Family, Betulaceae

Birches are moisture-loving trees of cold climates. Many are pioneer species, among the first to colonize bare ground. Most species are fast growing and relatively short-lived, although there are some exceptions, such as the yellow birch, which can grow to a venerable old age.

The paper birch, an iconic tree of the north woods, has bright yellow fall color.

PHOTO BY PETER M. DZIUK, www.MinnesotaWildflowers.info

Three species are prized for their colorful bark when young, which peels off in thin curly sheets as the trees expand in girth. The bark of river birch (*Betula nigra*) is a handsome salmon color. Paper birch (*B. papyrifera*) has bark that peels in striking chalky-white layers, while the bark of yellow birch (*B. alleghaniensis*) curls in shiny golden-bronze layers. All three of these popular species will be highlighted in this chapter.

Birch trees have male and female flowers, arranged in catkins, which occur on the same tree. The male catkins hang from the branch tips and lengthen in spring, when they release their pollen. The female catkins, only about one-half to one inch long, occur on short spur-like branches. After receiving the wind-borne pollen, they produce seeds that become tightly packed within their catkins.

Value to Birds and Other Wildlife

Birch trees are a one-stop supermarket for birds, stocked with seeds, sap, and an impressive variety of insect delicacies. American Goldfinches, Common Redpolls, Pine Siskins, Purple Finches, Red-breasted Nuthatches, Red Crossbills, and Ruffed Grouse are particularly fond of birch seeds, and in the fall and winter, these birds are commonly seen foraging for the seeds hidden in the catkins or fallen on the snow. American Goldfinches, Purple Finches, and Red Crossbills also eat the buds of birch trees, while Ruffed Grouse feed on seeds, buds, and catkins. Other birds that occasionally feed on birch seeds include the American Tree Sparrow, Black-capped Chickadee, Blue Jay, Cedar Waxwing, Dark-eyed Junco, Green-winged Teal, Fox Sparrow, Ring-necked Pheasant, Rose-breasted Grosbeak, Sharp-tailed Grouse, Tufted Titmouse, White-breasted Nuthatch, Wild Turkey, and Wood Duck. Yellow-bellied Sapsuckers drill holes in birches and feed on the oozing sap, which sometimes has a higher sugar content than that of maples. Hummingbirds visit to dine at the sap wells, a sweet addition to their diet of flower nectar. Sapsuckers also create and nest in cavities in birch trees. These cavities, along with naturally occurring cavities, are later used by the Black-capped Chickadee, Downy Woodpecker, Hairy Woodpecker, Tufted Titmouse, and White-breasted Nuthatch. Other birds build their nests in the leafy boughs of birches and use birch bark as a nesting material, including the American Redstart, American Robin, Baltimore Oriole, Black-throated Green Warbler, Blue-eyed Vireo, Philadelphia Vireo, Scarlet Tanager, White-eyed Vireo, and White-Winged Crossbill.

Perhaps the most important value of birches for birds are the hundreds of insects that are drawn to birch trees. Among woody species, birch hosts the fourth highest number of butterfly and moth species, more than four hundred! Caterpillars of the handsome mourning cloak butterfly feed on birch, along with those of the beautiful cecropia, io, luna, and promethea moths. Some common moths found on birch carry the tree's name, such as the birch dagger moth and the masked birch caterpillar moth. The latter, along with the chocolate prominent moth, eats only birch trees. A great variety of other insects, including aphids, beetles, bugs, leafhoppers, treehoppers, and weevils, also feed on birches. Many are so frequently associated with birch that they, too, are named after the tree: the birch lace bug, birch catkin bug, birch girdler, birch sawfly, bronze birch borer, dusky birch sawfly, European birch aphid, and fringed moth sawfly. All the insects, especially the caterpillars, are essential foods for the nestlings of many species of birds.

Some mammals also rely on birch for sustenance. Red squirrels eat the seeds. Beavers gnaw on the wood. Deer and rabbits browse on the twigs and leaves of saplings and seedlings. Bats sometimes use birches for roosting, breeding, and hibernation.

How to Grow

Yellow birch and paper birch, in particular, are very intolerant of drought and summer heat and are not recommended for urban areas but are good choices for cool climates with moist soils. River birch, however, is more adaptable and does well when grown in moist spring soils that may be drier in summer—typical of trees native to floodplains. This makes it the best choice for most home landscapes.

If pruning is needed, do so in fall or winter to avoid messy sap bleeding from the wounds. Older trees whose bark has darkened with age can be cut to the ground every twenty years or so. They will regrow readily, and the new bark of the young trees will again have colorful bark.

Although most insects attracted to birch are harmless and kept in check by birds and beneficial wasps, a few species can cause significant disease and even death of birches, especially in trees weakened by drought and heat. One of the worst is the bronze birch borer (*Agrilus anxius*), which is a native wood-boring beetle that can girdle the trees, but healthy trees are seldom harmed.

Redpolls subsist almost entirely on birch seeds in winter.
PHOTO BY JEREMY MEYER, jeremymeyer.photography

River birch is the birch most resistant to this pest. The birch leaf miner (*Funusa pusilla*) is a European sawfly that causes leaf drop. Paper birch is most susceptible to this pest. It does not usually kill the trees but can weaken them and make them more susceptible to other insects and diseases. There has been some success in controlling this pest with parasitic wasps.

Propagation

Seeds; difficult from cuttings.

Companion Plants

Avoid tall plants that could block views of the beautiful exfoliating bark of birches. Some herbaceous species known to grow in moist soils near birches include Canada mayflower or wild lily of the valley (*Mainthemum canadense*), cardinal flower (*Lobelia cardinalis*), and royal fern (*Osmundo spectabilis*).

A Closer Look

More than any other finches, Redpolls subsist almost entirely on birch seeds in winter. Small, roly-poly birds with a red patch on their heads, they breed in the Arctic tundra and boreal forests. Large flocks migrate south in winter only when northern seed crops fail. During such irruptive years, large flocks of Redpolls can show up feeding on birches and at thistle feeders as far south as the central United States. These amazing birds can store up to two grams (0.07 oz.) of seeds in pouches in their throats, equivalent to a fourth of their daily energy needs. They feed on birches in three phases: First, they knock the seeds from the cones, usually to snow-covered ground. Next, they gather up seeds and store them temporarily in their throat pouches. In the final phase, they fly off to a sheltered location protected from storms and predators, where they crack open the shells and eat the seeds. Keep a close watch for these lively, colorful birds that may visit occasionally some winters, as they have in my yard.

Yellow Birch (*Betula alleghaniensis*)

Deciduous Tree
Size: Height 60–80 feet, width 15–30 feet
Color and Season: Greenish flowers in spring; golden-yellow foliage in autumn
Light Requirement: Sun, partial shade
Soil: Moist, strongly acid to alkaline
Native Habitat: Wet to moist woodlands
Native to: CT DE GA IA IN KY MA MD ME MI MN NC NH NJ NY OH PA RI TN VA VT WI WV; rare in IL SC
USDA Hardiness Zones: 3–7

A tree of northern hardwood forests, yellow birch is a preferred nest tree for the beautiful red-shouldered hawk. This birch is a hardy and long-lived species, with old-growth specimens reaching three hundred years of age or more. Its bark peels in tight curls resembling rolled-up scrolls and has a distinctive glowing golden-bronze color on younger trunks and branches. Its sap and crushed twigs have a wintergreen aroma and taste. In late summer to early fall, it produces copious amounts of seeds, enticing many seed-eating birds to visit. The tree is the primary birch species used for lumber to make furniture and cabinetry. As a landscape tree, it is best grown on large sites in cool climates. In forested areas, it is rather narrow in habit, but in the open situations, it develops a spectacular wide-spreading crown. Along with its slender graceful branching and glowing fall foliage, yellow birch is an outstanding choice for a specimen tree.

River Birch (*Betula nigra*)

Deciduous Tree
Size: Height 40–75 feet, width 15–25 feet
Color and Season: Light green flowers in spring; yellow foliage in autumn
Light Requirement: Sun, partial sun
Soil: Moist to medium dry; prefers acid
Native Habitat: Floodplains along rivers and streams; sandbars
Native to: AL AR CT DE FL GA IA IL IN KS KY LA MA MD MI MN MO MS NC NJ OH OK PA RI SC TN TX VA WI WV; rare in NH NY
USDA Hardiness Zones: 4–9

A handsome, hardy, and fast-growing species, river birch is native to most of the eastern half of the country. Because it has the most southerly native range of all birch species in the country, it is the top choice among birch for landscaping. It is resistant to the bronze birch borer, rarely harmed by leaf miners, and is fairly disease resistant. It tolerates seasonal flooding as well as summer drought. However, it does not do well in high alkaline soils (pH 6.5 or over), which cause chlorosis, or yellowing of the leaves. Some experts believe this may be due more to dry conditions than soil pH. Grown in moist, acidic soils, river birch is an excellent specimen tree, ideal for rain gardens or near a downspout. Its exfoliating salmon-colored bark is highly ornamental all seasons of the year. River birch has also been used in wetland and floodplain restorations.

American Robins nest in the boughs of the river birch.
PHOTO BY DAWN-MARIE STACCIA

Paper Birch (*Betula papyrifera*)

Deciduous Tree
Size: Height 50–75 feet, width 20–30 feet
Color and Season: Greenish-yellow flowers in spring; yellow foliage in autumn
Light Requirement: Sun, partial shade
Soil: Moist; prefers acid
Native Habitat: Moist woodlands
Native to: AK CT IA ID IL MA ME MI MN MT ND NE NH NY OR PA RI SD VT WA WI WY
USDA Hardiness Zones: 2–6

An iconic tree of the north woods, paper birch is one of the most attractive native trees, with graceful branching, a handsome airy crown, and showy chalk-white bark. Because of its beauty, it's been called "the lady of the woods," and because Native Americans used it to make canoes it is also called canoe birch. They stretched sheets of bark over a frame of white cedar, sewing it with thin tamarack roots and caulking holes and seams with the resin of pine or balsam fir. Lightweight and easily portaged, the canoes carried these first peoples across waterways throughout the country and were soon adopted by fur traders and settlers.

Although a popular tree for landscaping, paper birch requires a cool climate and cool soil temperatures, conditions that are often difficult to provide, especially at the southern end of its native range in urban areas. When its environmental needs are not met, it is very susceptible to the bronze birch borer, which can destroy the tree. Avoid planting it near dry pavements or on sunny open lawns. Choose cool, moist locations where its roots are shaded, and mulch heavily to maintain soil moisture. If you don't have a suitable place, the river birch can be a good alternate choice.

HICKORY (*Carya* species)

Walnut Family, Juglandaceae

Hickories are major components of oak-hickory forests, one of the most extensive deciduous forest types in eastern North America. They replaced chestnuts as the major nut tree for wildlife after the chestnut blight wiped out chestnuts at the turn of the twentieth century. Some hickories are also found in savannas, where their tall straight trunks, broad canopies, and upturned branching lend them an attractive picturesque appearance, suggesting their use as a specimen tree for large properties. The genus name *Carya* recalls the ancient Greek nut tree goddess, *Carya*, a virgin who was transformed into a nut tree. The common name, hickory, is a shortened version of the Algonquian word *pockerchicory* or similar word.

There are twenty-one hickory species and naturally occurring hybrids in North America. The two most widespread and common hickories are shagbark

Shagbark hickories offer a bonanza of nuts, insects, and sap for birds.
PHOTO BY MARIETTE NOWAK

hickory (*C. ovata*) and bitternut hickory (*C. cordiformis*), both of which will be featured here. A natural hybrid between the two is Laney's hickory (*C. x laneyi*), which occurs in Wisconsin and a few northeastern states. All three are particularly showy in fall when their beautiful golden-yellow foliage is on display.

Hickories have both male and female flowers on the same tree and are wind pollinated. The male flowers are in drooping catkins up to six inches long; in contrast, the female flowers are tiny, only about a quarter inch long. Although they can self-pollinate, the trees produce more nuts if pollinated by other hickory trees.

Value to Birds and Other Wildlife

Hickories offer a bonanza of nuts, insects, and sap for birds. Although the nuts are too big for most birds' bills, the nuts sometimes crack open naturally or are chewed open by squirrels and other critters, giving birds access to the nutmeats. The species that are most fond of hickory nuts are the Carolina Chickadee, Eastern Towhee, Field Sparrow, Northern Cardinal, Pine Warbler, Red-bellied Woodpecker, Tufted Titmouse, White-breasted Nuthatch, and Yellow-rumped Warbler. Other birds also enjoy an occasional nibble, including the American Crow, Blue Jay, Mallard, Northern Bobwhite, Pine Grosbeak, Ring-necked Pheasant, Rose-breasted Grosbeak, Wild Turkey, and Wood Duck. The sugary sap of hickories is the food of choice for the Yellow-bellied Sapsucker, a woodpecker that drills a series of holes in the trunks of hickories (and other trees) and feasts on the upwelling sap. The Brown Creeper finds a cozy place to build its hammock-style nest under flaps of hickory bark, usually in a dead or dying tree. Other

A Red-headed Woodpecker in a hickory tree scans for insects, its major food in summer.
PHOTO BY DAWN-MARIE STACCIA

birds build their nests among the branches of hickories, including the Blue-gray Gnatcatcher, Cerulean Warbler, Eastern Wood Pewee, Prairie Warbler, Scarlet Tanager, and Yellow-throated Vireo.

The amazing assortment of insects on various species of hickory trees is sure to attract even more birds than do the nuts. Two hundred species of moth and butterfly caterpillars feed on hickories. Many additional insects can be found eating hickory leaves, forming galls on leaf stems, boring in the wood or bark, sucking plant juices, or eating the nutmeats. In turn, all these invertebrates are potential food for birds, who raise their nestlings on the diverse smorgasbord of these protein tidbits and, in doing so, help to reduce pests that may harm the trees. A number of these insects are so commonly found on hickories that they are named for the tree, notably the caterpillars of the hickory hairstreak butterfly and the hickory tussock moth. Even more beetles bear the name of their favorite tree, including the banded hickory borer, hickory bark beetle, hickory saperda, hickory twig pruner, living hickory borer, and painted hickory borer. Then there are weevils, such as the hickory nut curculio and the hickory shoot curculio. Each gall-making aphid on hickories creates a distinctive-shaped gall. Several species of ants glean sap beneath the bud scales. Another ant species builds huge nests within the rotting wood of standing trees.

Many mammals devour shagbark hickory nuts but avoid the extremely bitter nuts of the aptly named bitternut hickory, whose nuts are eaten only when nothing else is available. Bears, chipmunks, deer, fox, mice, raccoons, and squirrels—all eat the nuts of hickories with sweet meats. Deer also browse on the leaves and branches, while rabbits browse on seedlings. In winter, meadow voles gnaw on the tender bark of saplings. Beavers feed on the wood and use the branches to build their dams and lodges. In summer, bats roost in the crevices of the peeling bark, while cavities in the trees offer shelter and nest sites for other mammals.

How to Grow

Hickories have deep taproots, which make them difficult to transplant, but the trees are nonetheless available at native nurseries that have developed techniques to create more shallow, fibrous roots. Some experts recommend planting the nuts instead. Remove the plants around them to encourage better growth, and after the first growing season, cut the sprout back to ground level. The stems growing the following year will be more vigorous and better formed. For the best pollination and nut production, plant several trees or nuts to form a small grove of trees as they mature.

Propagation

Seeds (nuts).

Companion Plants

Some beautiful understory plants that are easy to grow and are found in the wild under hickory trees are common oak sedge (*Carex pensylvanica*), elm-leaved goldenrod (*Solidago ulmifolia*), Short's aster (*Symphyotrichum shortii*), and woodland sunflower (*Helianthus strumosus*).

A Closer Look

Brown Creepers not only nest within the furrows of shagbark hickory trees; they also forage for insects in the folds and pockets where insects are often especially abundant. Creepers are so small and well camouflaged that it's always a feat to spot them. When you do, it's fascinating to watch them as they creep up a tree trunk, searching for critters in the bark, and then fly down to the base of another tree to begin another long climb. In winter, creepers are particularly common in oak-hickory forests. If you have good hearing, you may be able to find them by listening for their call notes, as I've done on quiet

Brown Creepers creep up trees, searching for insects hidden in the bark.

PHOTO BY JEREMY MEYER, jeremymeyer.photography

winter days. The call is tinkling and high pitched, lasting longer than the short call notes of many other birds, and it is given all seasons of the year.

In the Midwest, Brown Creepers nest "up north" in Michigan, Minnesota, and Wisconsin, but they also occasionally nest in states as far south as Missouri and Virginia. In breeding areas, listen for their delicate, high-pitched song, which has been likened to the phrase "trees, beautiful trees." It's a lovely little song that I've been lucky to hear in northern Wisconsin. Although the song carries well through the forest, it is easy to miss. Keep your ears open when you're visiting the creepers' nesting grounds.

Bitternut Hickory (*Carya cordiformis*)

Deciduous Tree
Size: Height 75–100 (160) feet, width 20–40 feet
Color and Season: Greenish-yellow to green flowers in spring; yellow to bronze foliage in autumn
Light Requirement: Partial shade
Soil: Moist to medium dry; fertile loamy, but tolerates less fertile gravelly soil
Native Habitat: Woodlands
Native to: AL AR CT FL GA IA IL IN KS KY LA MD MA MI MN MO MS NE NH NJ NY NC OH OK PA RI SC TN TX VA VT WI WV; rare in ME
USDA Hardiness Zones: 4–9

Bitternut hickory has the widest range of any other hickory species and grows farther north. Its native habitat is primarily in river valleys and wooded bottomlands, but it may also be found in drier uplands. Try eating a nut and you won't forget its name. It is also called yellow-bud hickory for its bright yellow buds, which help distinguish it from other hickories and, indeed, from most other tree species. Bitternut hickory has a vase-shaped form, resembling the elm trees of old, and its bark is smooth and gray. It does not have the character and ornamental characteristics of shagbark hickory and is not as widely used in landscaping. But it is a species worth adding for diversity or to preserve if you're lucky to have one on your property.

Shagbark Hickory (*Carya ovata*)

Deciduous Tree
Size: Height 75–100 feet, width 20–35 feet
Color and Season: Yellowish-green or green flowers in spring; yellow to bronze foliage in autumn
Light Requirement: Partial shade, sun
Soil: Medium to medium dry; well-drained deep loam or clay loam
Native Habitat: Woodlands, savannas
Native to: AL AR CT GA IA IL IN KS KY LA MA MD ME MI MN MO MS NC NE NH NJ NY OH OK PA RI SC TN TX VA VT WI WV; rare in DE
USDA Hardiness Zones: 4–8

Shagbark hickory is a stately tree that can grow to immense girth and height and is long-lived, surviving for up to three hundred years. Its shaggy, peeling bark adds interest every season of the year, and it has one of the most ornamental budding of any tree, which should not be missed.

The inner bud scales enlarge in spring, becoming a lovely pinkish red, and arch down like a candleholder surrounding the emerging pinkish-red leaves, which gradually turn a contrasting fresh green color. Although the tree prefers moist, well-drained soils, its deep taproots enable it to grow on dry upland slopes along with oaks. Use this species as an excellent landscape tree for large open

The pinkish-red bud scales of hickories arch down like candleholders for the newly emerging leaves in spring.
PHOTO BY MARIETTE NOWAK

sites or in woodland restorations. Even though it hosts many insects, the critters are mostly unnoticed without close inspection and seldom harm healthy trees.

AMERICAN BEECH (*Fagus grandifolia*)

Beech Family, Fagaceae
Deciduous Tree
Size: Height 50–70 feet, width 25–50 feet
Color and Season: Yellow-green flowers in spring; bronze foliage in autumn
Light Requirement: Shade, partial shade
Soil: Moist; rich, slightly acidic loam
Native Habitat: Rich woodlands and forests
Native to: AL AR CT DE FL GA IL IN KY LA MA MD ME MI MO MS NC NH NJ NY OH PA RI SC TN TX VT VA WI WV; native but rare in OK
USDA Hardiness Zones: 3–9

Beech trees are the aristocrats of the forest. Tall and stately, they are easily recognized by their bluish-gray bark, which remains smooth even as they age, so different from the darker grooved, corrugated, or gnarly bark of most other trees, above which the beech often towers in the forest canopy. One of the most shade tolerant trees, beech frequently form groves, spreading by multiple shoots arising from their shallow roots. The trees are beautiful through every season. Winter showcases their silvery bark, distinctive chestnut-brown buds, and persistent foliage. Spring brings emerging chartreuse leaves, the color a striking contrast to the silvery bark. By summer, the tree has a dense verdant crown with horizontal branching that often sweeps gracefully toward the ground. By fall the golden-bronze foliage glows in the softer light of the season amid the silvery-barked branches.

The name *beech* is very similar and sometimes identical to the word *book* in many European languages. In antiquity, beech bark was often used as a writing surface before paper was developed, and beechwood tablets were common in Germany. Today, beech trees are sometimes called autograph trees because their bark is so frequently initialized by lovers and others. As the authors of *Flora of the Chicago Region* remark, “Many young couples have established their ‘beech-head’ on the trees.”

American beech trees have smooth bluish-gray bark, making them easy to identify.

PHOTO BY CLAIRE SECRIST, CC0 1.0, https://bplant.org/blog/8

Value to Birds and Other Wildlife

From nuts to caterpillars, bugs and beetles, American beech offers top dining for birds.

The fruits of beech trees are small triangular nuts contained in a bur-like husk with weak spines. The nuts are a favorite of many birds, including the American Crow, Blue Jay, Common Grackle, Northern Bobwhite, Northern Flicker, Pileated Woodpecker, Purple Finch. Red-bellied Woodpecker, Red-headed Woodpecker, Ruffed Grouse, Tufted Titmouse, White-breasted Nuthatch, Wild Turkey, and Wood Duck. The Purple Finch likes to munch on the buds as well, while the White-throated Sparrow feeds on the flowers. Most woodpeckers feed on the nuts, but the Yellow-bellied Sapsucker, a woodpecker by another name, prefers the sap.

Beech trees also provide cover and nesting sites for birds. Red-shouldered Hawks favor large trees like beech that stand above other tree species in forests for their nests. Songbirds such as the Acadian Flycatcher and Wood Thrush nest on horizontal forks in beech trees, and the thrush also uses fallen beech leaves to line its nest. Natural cavities in beech trees are used by chickadees, nuthatches, titmice, and woodpeckers.

The American beech, the only member of its genus in North America, supports more than 125 species of moth and butterfly caterpillars. Entomologist Doug Tallamy reports, "On a per-species basis, this makes beech more productive in its support of wildlife than even the number-one-ranked oaks." These caterpillars are prime food for nestling birds. Among the moths are several of our largest and most spectacular species, such as the lime-green luna moth with graceful, curving hindwing tails and the handsome cecropia moth with its rich reddish and tan patterning. Beetles and leafhoppers are also abundant on beech trees. Other insects that feed on beech include aphids, scale insects, and true bugs.

Mammals, unsurprisingly, also enjoy the bounty of beech trees. Bears, chipmunks, fox, deer, mice, and squirrels feast on beech nuts. Squirrels and other mammals den in cavities in the tree.

How to Grow

Beech trees grow well in woodlands with oaks and hickories and also make magnificent specimen trees on large properties. Plant in rich, moist soil in part shade and keep young trees well watered and protected in winter with fencing or a deer guard. Beech trees can be pruned for structure during the dormant season, but be very careful not to harm the thin bark or the roots, which may create openings for diseases. Anecdotal evidence suggests that beech bark disease, which is fatal to beech trees, is more common on trees with damaged bark due to engraving.

Propagation

Seeds (nuts).

Companion Plants

Beechdrops (*Epifagus virginiana*) are parasitic plants that grow only under American beech trees. They have tiny white-and-purple flowers and many brown stems with scalelike leaves. Having no chlorophyll, they survive by extracting nutrients from the shallow roots of the trees. Beechdrops are not purposely planted but cause no real harm and are interesting to observe.

Other than beechdrops, few plants can thrive below the heavy shade of beech trees. Moss, however, can be introduced as an attractive ground cover. Purchase moss or gather some from a nearby shady location (be sure to get permission if the moss is not on your own property). You do not need much; simply press the moss into the soil in evenly spaced patches. Be careful to avoid harming the shallow roots of the trees. The moss will spread on its own in time.

A Closer Look

> The pigeon loved his land: he lived by the intensity of his desire for clustered grape and bursting beechnut, and by his contempt of miles and seasons.
>
> —Aldo Leopold, *On a Monument to the Passenger Pigeon*

Beechnuts were the favorite food of the extinct Passenger Pigeon, immortalized here in this painting by John J. Audubon.

DRAWING BY JOHN J. AUDUBON, WIKIMEDIA COMMONS, https://commons.wikimedia.org/wiki/File:Passenger_Pigeon_Audubon.jpg

Beechnuts were the Passenger Pigeon's favorite food. These beautiful birds, once numbering in the billions, darkened the skies for miles during migration and took days to pass. They often chose to nest en masse in forests where beech trees were abundant. Beechnuts are extremely nutritious nuts and contain more than 40 percent fat, helping to feed the pigeons' nestlings and fuel their vast migrations. Despite the pigeons' staggering numbers, they became extinct due to overhunting and loss of habitat, as the large forests needed for nesting were lost to lumbering and farming. A monument to the Passenger Pigeon was placed in Wyalusing State Park by the Wisconsin Society for Ornithology and dedicated on May 12, 1947, reminding us to never forget the loss of this magnificent bird and to work tirelessly to end the continuing loss of species in this time of the Sixth Extinction.

CRABAPPLE (*Malus* species)

Rose Family, Rosaceae
Deciduous Tree
Size: Height 15–20 feet (some 35 feet), width 15–20 feet
Color and Season: White, pink, magenta, or red flowers in spring; varied colored foliage in autumn
Light Requirement: Sun
Soil: Moist to moderately dry; adaptable to variety of soils
USDA Hardiness Zones: 3–9

Several species of crabapple are native to the United States. Unfortunately, all have fruits that are too large for most birds, usually about three-quarters to one inch or more in diameter. In contrast, the fruit of the non-native ornamental flowering crabs that are so popular at nurseries are often the smaller size that birds prefer. Our native insects thrive on these non-native crabs as well. Entomologist Douglas Tallamy explains that ornamental crabs are apparently "so similar in leaf chemistry to our native species that there is little evidence that native insects can tell the difference, particularly when these plants hybridize." This is one case where non-native plants are equally attractive to our insects, and those with smaller berries provide better fruit than native species for our birds.

Warblers, like this Golden-winged Warbler, flock to flowering crabapples to gorge on the many insects attracted to the blossoms.

PHOTO BY JEREMY MEYER, jeremymeyer.photography

Crabapples are small trees that explode with colorful, fragrant flowers in spring and produce fruit prolifically in late summer to mid-fall, with some holding their fruit through winter. Many also have bright fall foliage. There are nearly one thousand varieties of crabs from which to choose. Those best for birds will have berries no more than about three-quarters of an inch in width, a perfect size for most birds' beaks. Crabs whose fruit persists though fall and perhaps even through winter will attract more birds than those that drop their berries quickly. Crabapple flowers with a single circle of petals will ensure a good supply of insects for birds, in contrast to those having multiple layers of petals that are difficult for pollinators to access. Another important feature to consider is resistance to apple scab, cedar apple rust, and fire blight–diseases to which crabs are particularly susceptible.

Most crabs have fruit that is initially hard and marble-like. As they gradually soften after freezing and thawing several times through fall and winter, they become palatable to birds. The fruit of various cultivars soften at different rates. Horticulturist Mike Yanny's excellent article "Crabapples for Birds," on the website of Johnson's Nursery in Wisconsin, describes the characteristics of many varieties of crabapples. The fruits of some crabs soften quickly in mid to late autumn, while others soften in late winter and may even become fermented in the process, providing a bit of booze for the birds. There are also some crabs with small fruits that birds never eat for some unknown reason and others that fruit only every other year, both of which you'll want to avoid. There are even a few crabapples that never produce fruits at all, apparently grown primarily for their flowers. Obviously, you have many things to consider before selecting a crabapple, but it will be worth the research when birds flock to your yard through the seasons.

Value to Birds and Other Wildlife

Crabapple trees provide a smorgasbord for birds through the year. In spring, many birds visit to feed on the flowers, nectar, and sap. Ruby-throated Hummingbirds come to the flowers to dine on nectar. Cedar Waxwings and Purple Finches are particularly fond of the flower blossoms and later in the year eat the berries too. Many birds also visit the blossoms to dine on the insects that come for nectar and pollen. More than three hundred butterfly and moth species feed on various parts of crabs and offer tasty morsels for birds. Yellow-bellied Sapsuckers drill for sap, which, in turn, entices hummingbirds to visit for a sip of sap and to snag any insects also attracted to the sap. The fall and winter berries of various crabapple species are relished by a wide variety of birds, including the American Crow, American Robin, Baltimore Oriole, Blue Jay, Canada Jay, Common Grackle, Eastern Towhee, *European Starling*, Gray Catbird, Northern Cardinal, Mockingbird, Orchard Oriole, Red Crossbill, Ruffed Grouse, Tufted Titmouse, Thrushes, Waxwings, and most woodpeckers, including Downy, Hairy, Red-bellied, and Red-headed, as well as the Yellow-bellied Sapsucker. Crabapple

A Hermit Thrush dines on crabapple fruits before it sets out for its southern wintering grounds.
PHOTO BY JACK BARTHOLMAI

buds as well as berries are eaten by Northern Bobwhite, Ring-necked Pheasant, and both Evening and Pine Grosbeaks. Many birds use the trees for cover and nesting sites, including a few, like bluebirds, vireos, kingbirds, and flycatchers, that haven't been recorded eating the berries.

How to Grow

Crabapple trees grow best in full sun, and many are quite adaptable to a variety of soil conditions and even urban pollution. As mentioned, choose those resistant to disease. Consider planting at least two crabapples for better pollination. Since the fruits of different cultivars tend to soften at different times throughout the fall to early spring, you may also want to plant a mix of cultivars to provide for the birds through the seasons. It can take two to five years before crabs begin to produce berries.

Propagation

Native crabapples: seeds. Ornamental crabs: budding, grafting, cuttings, or tissue culture.

Companion Plants

A low ground cover of common oak sedge (*Carex pensylvanica*) offers a lovely green carpet to set off these attractive trees.

A Closer Look

Crabapple birding is a spring highlight for many birders. There's nothing more beautiful than birds amid blossoms. Wood Warblers, in particular, seem to be attracted to these flowering trees and bounce about here and there as they feed on insects. Feathered in a rainbow array of colors, these beautiful birds are often referred to as the butterflies of the bird world. Bird-watchers are often able to get awesome views of these warblers perching in sunny openings in the foliage, surrounded by petals in pastel shades of white and pink. American Redstart, Blue-winged Warblers, Golden-winged Warblers, Magnolia Warblers, Yellow Warblers, and more—their names often revealing their colorful plumages—are among the many species my friends and I have discovered in spring-flowering crabapple trees. Don't miss the show.

WHITE SPRUCE (*Picea glauca*)

Pine Family, Pinaceae
Evergreen Tree
Size: Height 40–60 feet, width 10–20 feet
Color and Season: Yellow male flowers and purple female flowers in spring
Light Requirement: Sun, part sun
Soil: Moist; tolerant of many soils but not constant flooding
Native Habitat: Boreal forest
Native to: AK ME MI MN MT NH NY SD VT WI WY; rare in ID
USDA Hardiness Zones: 2–6

White spruce is the northernmost tree species in North America and a major tree in boreal forests. Like all spruces, it has a spire-like form with dark evergreen, four-sided needles with sharp tips—the classic Christmas tree shape. The cones, smallest among the spruces, hang down from the branches, unlike the upright cones of firs. A hardy, adaptable tree, it grows well in moist, cool forests and has a lifespan of 250 to 300 years. White spruce is one of two common spruce species native to the Upper Midwest and Northeast. The other is black spruce (*Picea mariana*), found primarily in northern bogs and swamps, for which reason it is rarely used in landscaping. Another native species, red spruce (*P. rubens*), is restricted to mountaintops in the Northeast from Canada and Maine south to North Carolina. Several spruces non-native to the eastern half of the country are nonetheless widely used in landscaping, including the Norway

Cedar Waxwings are among the many birds that nest in the white spruce.
PHOTO BY JACK BARTHOLMAI

spruce (*P. abies*), native to Europe, and the Colorado blue spruce (*P. pungens*), native to our western states.

Value to Birds and Other Wildlife

White spruce trees nourish birds with their seeds, sap, and needles, plus the many insects they host. In spring and summer, warblers and many other birds forage for their caterpillars and other insects. In a classic study, renowned ecologist Robert MacArthur recorded the feeding habits of different warbler species in spruce trees. Cape May warblers fed in the outer branches with young needles at the top. Bay-breasted Warblers fed primarily in the middle interior, while Yellow-rumped Warblers were more versatile and used various sections of the trees. They foraged for insects differently too. Some caught insects on the fly; others more often hovered. In addition, the warblers had variable nesting times, so their peak need for food was staggered. Each warbler species, the research revealed, had its own unique niche within the same spruce trees, helping to avoid over-competition for the same food source at the same time.

Among the many insects supporting warblers and other birds are more than 150 species of butterfly and moth caterpillars. Not surprisingly, birds often have outstanding breeding success when there are outbreaks of caterpillars like the spruce budworm. Other insects that feed on spruce include the yellow-headed spruce sawfly, European spruce sawfly, spruce bud moth, spruce beetle, gall formers, bud midges, leaf miners, aphids, leaf eaters, leaf rollers, loopers, mites, scales, weevils, borers, pitch moths, and spittlebugs—all morsels for warblers and other birds.

Spruce seeds ripen in late summer, persisting well into winter, and carry many birds through the winter. Birds that specialize in feeding on spruce seeds are primarily northern birds, including the Evening Grosbeak, Pine Siskin, Purple Finch, Red-breasted Nuthatch, Red Crossbill, White-throated Sparrow, and White-winged Crossbill. Many other birds snack on the seeds on occasion: American Goldfinch, Black-capped Chickadee, Boreal Chickadee, Cedar Waxwing, Chipping Sparrow, Downy Woodpecker, Hairy Woodpecker, Mallard, Mourning Dove, Pileated Woodpecker, Pine Grosbeak, Swainson's Thrush, and Wood Thrush. The Spruce Grouse, named after the trees, eats the needles of spruce and pine almost exclusively in the winter months, although some experts report that it eats the spruce seeds as well. The Yellow-bellied Sapsucker prefers drilling for the sap.

Besides warblers, many other bird species nest in spruce trees. Chickadees and woodpeckers bring up their nestlings within the trees' cavities. The dense needled boughs of spruce are a favorite nesting site for songbirds like American Robins, Chipping Sparrows, Mourning Doves, Northern Mockingbirds, and Cedar Waxwings. The dense cover also provides shelter for many birds during stormy and wintry weather.

Like birds, many mammals, including chipmunks, mice, shrews, squirrels, and voles, feed on the spruce seeds. The seeds are the primary food for red squirrels, which harvest and cache the cones and can live all winter on nothing but the nutritious seeds. Red squirrel populations can rise and fall with the seed crop. Snowshoe hares eat the twigs, buds, and bark of spruce trees.

How to Grow

With its dense needled foliage, white spruce is an excellent choice for a windbreak or a screen to hide an unsightly view. It also excels as a specimen tree. Although fairly adaptable, spruce does best in full sun, cool temperatures, and moist soils. A previous owner planted white spruce trees on my property in a shady location on dry soil where they're doing very poorly; the trees look thin and scraggly, and I'm gradually removing them. Spruce are also sensitive to heavy pollution and salt spray so it's best not to plant them along roads. In suitable landscape conditions, spruce grow well with minimal insect problems.

Propagation

Seeds.

Companion Plants

Few plants can grow directly under the heavy shade of spruce trees, whose low branches often sweep the ground. Various species of mosses are the most common plants found under the trees and can be introduced, as described for the ground cover under beech trees. Other plants that often grow near spruce in the

White-Winged Crossbills have crossed bills that help them extract seeds from spruce cones.
PHOTO BY JACK BARTHOLMAI

wild include blueberries (*Vaccinium* species), bunchberry (*Cornus canadensis*), and sedges (*Carex* species).

A Closer Look

White-winged and Red Crossbills are fascinating birds that specialize in feeding on the seeds of spruce and other coniferous trees and often nest in conifers. Their unique crisscrossed bills are an adaptation for extracting seeds deep inside conifer cones, which are not easily accessible to other birds. They use their twisted bills to wedge open the scales of the cones and then use their tongues to reach the seeds. They remove the seed coats before eating the seeds. If the cones are on the ground, they rotate the cones and eat them like corncobs. White-winged Crossbills can eat up to three thousand seeds a day.

Look for these birds in winter in stands of spruce on your own property or in cemeteries, parks, or forests. Both crossbills are stocky finches, a little bigger than the common house sparrow. Usually found in lively flocks, they travel south from the boreal forests of Canada to the northern tier of states and sometimes farther south, when food is scarce in their far north haunts. They chatter noisily and hang upside down in the trees like miniature parrots as they feed on the cones. You may hear distinctive crunching sounds as they split open the cones.

PINE (*Pinus* species)

Pine Family, Pinaceae

Pines are ancient plants that first appeared during the Mesozoic Era, the age of the dinosaurs. They are found throughout the Northern Hemisphere and have greatly diversified through time, with more than forty native species in the United States and Canada. But only three are common in the deciduous forest areas of the Midwest and northeastern United States and will be featured here. The one most widely used in landscaping is the majestic white pine (*Pinus strobus*). Jack pine (*P. banksiana*) is a small scraggly tree and less ornamental but an important species for wildlife, especially for the Kirtland's Warbler, one of the rarest songbirds in North America. Red pine (*P. resinosa*), named for its reddish bark, has the most northerly range of the three.

Value to Birds and Other Wildlife

Pines are one of the premier genera of trees for birds and other wildlife, ranked second only to oaks. Their seeds and needles, plus the many caterpillars and insects found feeding on pines, provide a nutritious menu for birds. Several birds

An acrobatic Red-breasted Nuthatch forages on a white pine.
PHOTO BY JACK BARTHOLMAI

are named after the trees they feast on, including the Pine Grosbeak, Pine Siskin, and Pine Warbler. The Pine Warbler is particularly well named because it eats more pine seeds than any other wood warbler, almost always nests in pine trees, and specializes in foraging for caterpillars on the foliage and bark of pine trees. Dozens of other birds show a preference for pine seeds, including both the Red Crossbill and White-winged Crossbill and three species of woodpeckers: the Hairy, Red-bellied, and Pileated. The Yellow-bellied Sapsucker sips on the sap. A group of small songbirds that often forage together are also known to prefer pine seeds: the Black-capped, Boreal, and Carolina Chickadees; the Brown-Headed, Red-breasted, and White-breasted Nuthatches; and the Brown Creeper. Other birds that prefer pine seeds include the Chipping Sparrow, Common Grackle, Dark-eyed Junco, Eastern Towhee, Evening Grosbeak, Mourning Dove, Northern Cardinal, Northern Bobwhite, and White-throated Sparrow. Spruce Grouse and Wild Turkey are also especially fond of pine seeds and eat the needles as well. Another dozen species feed on pine seeds occasionally, including the American Goldfinch, Carolina Wren, Cedar Waxwing, Common Redpoll, Eastern Meadowlark, Brown Thrasher, House Finch, *Monk Parakeet* (in urban areas), Purple Finch, Ring-necked Pheasant, Rock Dove, and Wood Duck.

A remarkable 203 species of moth and butterfly caterpillars grow fat on pine needles, adding another huge source of food for birds. A good number specialize in feeding on nothing else but pine, with some named after the trees, including the following moths: pine devil, northern pine sphinx, pine sphinx moth, and two loopers, the pine and northern pine. The white pine angle moth is even more exclusive, feeding only on white pine. The imperial moth feeds on a variety of trees but most commonly on pines. This fascinating moth has variable coloration, as both a caterpillar and an adult. Sawfly larvae, which look like caterpillars, can be found on most pine trees. Entomologist Doug Tallamy reports that these larvae can be invaluable to nesting Eastern Bluebirds during cold springs when few other insects are available to feed their young.

Not to be outdone are the other insects feeding on pines, some of whom are similarly named after the trees, including the flat-headed pine heartwood borer, jack pine resin midge, pine false webworm, pine root collar beetle, pine shoot beetle, pine needle scale, pine tree cricket, red-shouldered pine borer, and ribbed pine borer. All offer additional food choices for birds.

Pine plantations appear to be a prime nesting habitat for Cooper's Hawks in the Midwest.
PHOTO BY JEREMY MEYER, jeremymeyer.photography

The dense evergreen cover is a favored nesting site for some birds, like the Blue-headed Vireo, Pine Warbler, Yellow-throated Warbler, and Black-throated Green Warbler. Pine plantations appear to be a prime nesting habitat for Cooper's Hawks in the Midwest, especially in southeastern Wisconsin, where some of the highest recorded breeding densities have been discovered. Red-shouldered Hawks and Red-tailed Hawks also are frequently found in pines. Bald Eagles often construct their huge nests on tall super-canopy pines like white pine, which provide easy access and good visibility of their surroundings. The rare Kirtland's Warbler nests at the base of jack pine thickets. The dense needled foliage of pines also offers cover for many birds, including owls.

Among mammals, squirrels and mice feed on pine seeds. Deer and rabbits browse on the branches and needles.

How to Grow

Pines prefer sunny sites and acidic sandy soils and do poorly in clay or compacted soils. White and red pine are intolerant of salt spray from roads and in general are not good choices for urban areas. White pine blister rust, an introduced disease, causes powdery rust on the bark and can harm or even kill the pines. The alternate hosts are *Ribes* species, such as gooseberries and currants, which should not be planted near white pines. The *Ribes* species themselves are also harmed by the rust and will lose leaves and develop leaf spots if infected.

Don't be concerned about the seasonal needle drop in pines. In autumn, older needles in the center of the trees turn yellow, then brown, and drop off. This is a natural occurrence, similar to the annual leaf drop in deciduous trees, although in pines, the newer outer needles are retained for several years.

Propagation

Seeds.

Companion Plants

Good choices for herbaceous plants to grow with white and red pine include wild columbine (*Aquilegia canadensis*), sarsaparilla (*Aralia nudicaulis*), and blue-stemmed goldenrod (*Solidago caesia*). For jack pine, some species that associate with it in dry savannas are woodland sunflower (*Helianthus divaricatus*), western sunflower (*Helianthus occidentalis*), and Indian grass (*Panicum virgatum*).

A Closer Look

> Perhaps no bird species is more characteristic of the pine forests of eastern North America than the Pine Warbler.
>
> —W. R. Dawson, *Birds of the World*

Depending on the season, you'll likely find Pine Warblers wherever there's a stand of pines in your neighborhood. They are one of the few warblers that spend their entire lives in North America. They spend their winter in southeastern pines. In spring, some travel north to breed, passing through the lower Midwest during migration before they settle to breed in the Upper Midwest and Canada. Pine Warblers nest and forage high in the canopy where they are difficult to see. The best way to find them is to listen for their sweet song from high in pine trees. Their trilled song can be confused with that of the Chipping Sparrow and the Dark-eyed Junco, which have similar trilled repertoires. The Pine Warbler's song is the loveliest—more musical, slower, and softer than the other two. The Dark-eyed Junco's song is looser, less soft, and less musical. The Chipping Sparrow's song is a chipping rattle, considerably less musical than either of the other two. It takes careful listening to hear the differences, but check online sources like the Cornell Lab of Ornithology to practice. Once you

The well-named Pine Warbler nests and forages primarily in pines.

PHOTO BY JEREMY MEYER, jeremymeyer.photography

hear the Pine Warblers up in the trees, you may be able to catch glimpses of the birds in openings in the branches. You may also be lucky to see pine warblers at your feeders in winter and early spring. They are one of the few warblers that visit feeders and have been known to eat millet, cracked corn, sunflower seeds, peanuts, and suet.

Jack Pine (*Pinus banksiana*)

Evergreen Conifer
Size: Height 30–70 feet, width 10–25 feet
Color and Season: Flowers borne on cones in late spring; purplish female cones, yellowish male cones
Light Requirement: Sun
Soil: Moist to dry; well-drained acidic sandy soil
Native Habitat: Barrens, sand prairies, savannas, beaches; rock sandstone cliffs
Native to: ME MI, MN, ND, NJ, WI; rare in IA IL IN NH NY
USDA Hardiness Zones: 2–6

Jack pine is a rugged tree that thrives in poor soils, harsh winds, and scalding sun. It is a hallmark species of the sand counties of Wisconsin, described in Aldo Leopold's renowned book, *A Sand County Almanac*. The tree has an irregular asymmetric form and branching, which gives it a kind of unique bonsai appeal. Jack pine forests provide essential breeding habitat for the rare Kirtland's Warbler, which was on the federal endangered list for almost fifty years, until 2019, when their populations began to stabilize. Most of the warblers breed in Michigan, but a few also nest in Wisconsin and Ontario. The birds prefer to nest on the ground in the dense understory of very young jack pines, which are less than fifteen feet tall. As the trees grow and lose their lower branches, the warblers gradually abandon the site. The young jack pine forests were originally maintained by naturally occurring forest fires, but today timber harvests and prescribed burns are carried out to restore the young forests. The tree's cones open and release their seeds during burns and readily reseed the site.

Red Pine (*Pinus resinosa*)

Evergreen Conifer
Size: Height 50–80 feet, width 1/2 to 3/4 height
Color and Season: Flowers borne on cones in spring; purplish female cones, yellowish male cones
Light Requirement: Sun
Soil: Moist to dry; well-drained acidic soil
Native Habitat: Northern hardwood and boreal forests
Native to: MA ME MI MN NH NY PA VT WI; rare in CT IA IL NJ WV
USDA Hardiness Zones: 2–5

Red pine has sometimes been called Norway pine, a name mistakenly given to this American species by early English explorers who confused it with the European Norway spruce. A handsome tree with a straight trunk and a rounded picturesque crown, red pine requires the cooler climates of the Upper Midwest and northeastern states. In the 1930s, it was widely planted south of its natural range by the Civilian Conservation Corps, but the remaining trees are very prone to disease and insect problems in warmer, more humid conditions. Both the Morton Arboretum and the Missouri Botanical Garden no longer recommend planting red pine south of its normal range (generally south of USDA zone 5).

White Pine (*Pinus strobus*)

Evergreen Conifer
Size: Height 85–120 feet, width 60–90 feet
Color and Season: Flowers borne on cones in late spring; red to purplish female flowers and yellowish male flowers
Light Requirement: Sun to part sun
Soil: Moist to dry moist; well-drained acidic soil
Native Habitat: Forests, savannas, forested swamps and bogs
Native to: AL CT DE GA IA IL KY MA ME MD MI MN NC NH NJ NY OH PA RI SC TN VA VT WI WV; rare in IN
USDA Hardiness Zones: 3–8

The majestic white pine is the tallest conifer in the eastern half of the country and can live up to 350 years or longer. The classic pine of the north woods, it is also native to the South, if in lesser abundance. One of the best pines for native landscaping, it has handsome tiered branches with soft needles in bundles of five. The white pine is often planted as a specimen and shade tree in parks and lawns and has also been used in rows as a windbreak. The tree sometimes loses its leader in storms, giving it an irregular flat-topped look, but this only adds to its wild, windswept beauty. The chicken fat mushroom (*Suillus americanus*) sometimes shows up beneath white pine trees after a late summer rain. The mushroom sports a slimy yellow cap with reddish spots or stains. But not to worry. It actually has an important mycorrhizal association with white pines, providing the tree with nutrients from the soil, while receiving food in the form of carbohydrates from the pine—a win-win arrangement.

POPLAR (*Populus* species)

Willow Family, Salicaceae

All *Populus* species are short-lived pioneer species, often the first trees to sprout in disturbed habitats where they help heal the earth and prepare it for more longer-lived trees. One reason for their short life spans is their weak

Quaking aspens are particularly showy in fall with brilliant golden-yellow leaves.
PHOTO BY PETER M. DZIUK, www.MinnesotaWildflowers.info

wood and branch structure, which makes them susceptible to damage in storms. They all have oval or broadly triangular leaves with more or less flattened stems, causing their leaves to quiver and glitter in the lightest breezes. Male and female flowers occur in catkins on separate trees, with only female trees producing seeds within the catkins. All have yellow fall foliage, but quaking aspens are particularly showy with leaves like brilliant golden-yellow coins shimmering in the autumn winds. The three common *Populus* species native to the eastern half of the country are quaking aspen (*P. tremuloides*), large-toothed aspen (*P. grandidentata*), and eastern cottonwood (*P. deltoides*)—all featured here.

Value to Birds and Other Wildlife

Aspens and cottonwoods nourish birds with their buds and their catkins packed with seeds, as do the insects that feed on the trees and provide nutritious protein for both adult birds and their nestlings. Prairie Chickens, Ruffed Grouse, and Sharp-tailed Grouse feed on both the buds and catkins. The Ring-necked Pheasant is known to feed only on the catkins. Other birds feed only on the buds, including the Baltimore Oriole, Evening Grosbeak, Northern Shrike, Pine Grosbeak, Purple Finch, and Rose-breasted Grosbeak. The Yellow-bellied Sapsucker drills holes in the trunks to tap the tree sap.

Among the over 360 species of lepidoptera caterpillars harbored by *Populus* species are 7 species of giant silk moths, 7 species of sphinx moths, and 77 species of noctuid moths. Caterpillars of showy butterflies like the Compton tortoiseshell, mourning cloak, red-spotted purple, tiger swallowtail, viceroy, and white admiral also fatten up on poplars, as do many aphids, beetles, bugs, leafhoppers, sawflies, and weevils.

In addition, *Populus* trees provide leafy nest sites for birds, such as the American Robin, Baltimore Oriole, Blue Jay, Common Grackle, Eastern Kingbird, Mourning Dove, Northern Parula, Orchard Oriole, Warbling Vireo, and Yellow Warbler. As these relatively short-lived trees begin to deteriorate and die after forty-five to eighty years, largely due to wood-decaying fungi, they offer additional value for cavity-nesting birds. The shelf fungi seem to give woodpeckers a visual clue that the trees have a soft core for excavating cavities. Most woodpeckers drill new cavities each year, which, when vacated, are used by many other birds. Seven woodpecker species found east of the Mississippi River are the primary cavity nesters in *Populus* trees: Downy Woodpecker, Hairy Woodpecker, Northern Flicker, Pileated Woodpecker, Red-bellied Woodpecker, Red-headed Woodpecker, and Yellow-bellied Sapsucker. The secondary cavity nesters that use the woodpecker cavities the following years are the Black-capped Chickadee, Saw-whet Owls and Screech-Owls, Eastern Bluebirds, *European Starling*, House Wrens, and Red- and White-breasted Nuthatches.

Many mammals also use *Populus* trees for food and shelter. Deer and elk browse on the twigs and leaves. Rabbits and voles nibble on the bark and sprouts in winter. Some squirrels eat the buds and den in the trees' cavities. Beavers and muskrat eat the bark, foliage, and branches. The trees are also especially important to beavers for the construction of their dams and lodges.

How to Grow

Aspens and cottonwoods are among the easiest trees to grow. They all prefer moist soils and full sun and will rapidly gain two to six feet a year in height. As these rather short-lived trees decay and lose branches, they can be cut to the ground in winter and will resprout abundantly throughout the area. Healthy standing trees, with the exception of cottonwood, also send out multiple sprouts. To reduce sprouting, mulch to keep the soil cooler and avoid damaging the bark. Some sprouts will occur nonetheless and can be controlled by regularly mowing lawns. In my small pocket prairies, where they show up among prairie flowers and native grasses, I cut the sprouts back to the ground. Recently, I have tried dabbing the cut stems lightly with an herbicide, which hasn't seemed to affect the nearby mature trees to which they are likely to be connected by underground roots.

Propagation

Seeds.

Companion Plants

Poplars are sometimes planted with alders to stabilize banks, since alders fix nitrogen and some of it is used by poplars to spur faster growth. In home landscapes, herbaceous species can be used as ground covers beneath poplars. Some good choices commonly found under the trees in natural areas are oak sedge (*Carex pensylvanica*), wild geranium (*Geranium maculatum*), and woodland sunflower (*Helianthus divaricatus*).

A Closer Look

The Ruffed Grouse is more closely associated with aspens, particularly quaking aspen, than any other bird species. The aspen is one of the most widely distributed trees in North America, and, perhaps not surprisingly, the Ruffed Grouse has the widest distribution of any game bird. Ruffed Grouse are resident birds in all of Canada, in Alaska, and in thirty-seven of the lower forty-eight states. Gordon W. Guillion, a professor at the University of Minnesota who studied Ruffed Grouse for thirty-two years, suggested a coevolution of aspens and grouse. Aspens have many characteristics that benefit the grouse. Their fleshy flower buds on stout twigs offer important winter browse for grouse. In spring, these buds elongate into nutritious catkins just in time for grouse breeding. And aspens' prolific root sprouting creates large clonal colonies to shelter and feed the birds.

The Ruffed Grouse is grayish or reddish brown with intricate patterns, making them well camouflaged and difficult to see. But you can listen for the male grouse's signature drumming in spring in young aspen woods. The grouse usually stands atop a log or rock, puffs up its plumage, raises its black crest, and then beats its wings back and forth in the air, creating miniature sonic booms.

The Ruffed Grouse is more closely associated with aspens than any other bird species.
PHOTO BY JEREMY MEYER, jeremymeyer.photography

These deep thumping booms can be heard up to a quarter of a mile away. In winter, grouse often bury themselves in the snow, and you may get a glimpse—and a heart-stopping surprise—if one suddenly explodes to the surface as you ski or hike nearby.

Eastern Cottonwood (*Populus deltoides*)

Deciduous Tree
Size: Height 75–100 feet, width 25–50 feet
Color and Season: Red flowers in catkins in early spring; pale yellow foliage in autumn
Light Requirement: Sun
Soil: Moist; adaptable to variety of well-drained soils
Native Habitat: Floodplain forests and banks of streams, rivers, and lakes
Native to: All states except AK HI ME; native but rare in OR WA
USDA Hardiness Zones: 2–9

The largest tree in the Upper Midwest, eastern cottonwood is named for the tiny cottony seeds produced in the catkins of the female trees. The abundant seeds are carried long distances by the wind and can clog gutters, sewers, and air-conditioner filters. Cottonwoods also have spreading shallow roots and should not be planted near sidewalks, sewers, or water pipes. For these reasons,

The Warbling Vireo, a nondescript bird with a lilting song, nests in cottonwood trees.
PHOTO BY JACK BARTHOLMAI

this species is generally not recommended for residential areas. As one of the faster-growing trees, however, cottonwoods can help to quickly shade and reforest larger properties and restoration sites. Several cottonwood trees on my two-acre property are apparently male, since they don't produce seed. Nor do cottonwoods resprout and form large colonies, as do quaking aspens. To my delight, Baltimore Orioles often nest in the trees, and noisy Pileated Woodpeckers have pecked large oblong cavities in a few dead cottonwood snags in search of carpenter ants.

Large-Toothed or Big-Tooth Aspen (*Populus grandidentata*)

Deciduous Tree
Size: Height 30–40 feet, width 20–35 feet
Color and Season: Gray flowers in catkins in spring; golden-yellow foliage in autumn
Light Requirement: Sun
Soil: Moist to dry; sandy loam but adaptable to variety of soils
Native Habitat: Open woodlands
Native to: CT DE IA IL IN KY MA ME MI MN NH NJ NY OH PA RI TN VA VT WI WV; native but rare in MO NC SC
USDA Hardiness Zones: 3–6

Both the common name and species name, *grandidentata*, refer to the large teeth of this poplar species. Its leaves are similar to those of the quaking aspen, a close relative, but are rounder and slightly larger with larger teeth. Like a middle child among the three common *Populus* species, large-toothed aspen is neither quite as beautiful as quaking aspen nor as impressive as a huge mature cottonwood. It is also less widespread, occurring primarily in the Great Lakes region and northeastern states, while the other two are found across the continent. This species has deeper roots than the other two species, enabling it to grow on drier upland sites. And although it spreads by root suckers, like quaking aspen, it does so less frequently and forms smaller clones. Large-toothed aspen also grows faster than quaking aspen and is more disease resistant, making it an excellent choice for parks and other large sites, and for soil stabilization.

Quaking or Trembling Aspen (*Populus tremuloides*)

Deciduous Tree
Size: Height 35–50 feet, width 15–35 feet
Color and Season: Flowers in brownish catkins in early spring; golden foliage in autumn
Light Requirement: Sun
Soil: Moist to moderately dry; tolerant of variety of soils
Native Habitat: Woodlands, wetlands, disturbed sites
Native to: AK AZ CA CO CT DE IA ID IL IN KY MA MD ME MI MN MT ND NH NJ NM NY OH OR PA RI SD TX UT VT WA WI WV WY; native but rare in MO NC NE VA
USDA Hardiness Zones: 1–6

The leaves of quaking aspen aspen quake and flutter in the lightest of winds, hence the name. I enjoy sitting on my deck and listening to the nearby aspens rustling softly, like a gentle rain shower. Their leaves, lustrous green in summer, golden yellow in fall, sparkle in the sunlight as they catch the breezes. Quaking aspen's bark is usually greenish to whitish gray and smooth, except lower on the trunks of older trees. The greenish bark gives them the ability to photosynthesize even when they're leafless in fall and winter. The prolific root sprouting of quaking aspen can cause problems (see the section How to Grow in the *Populus* introduction for ways to reduce the sprouting). This characteristic has resulted in a huge clone in Utah that has been honored as one of the "40 Wonders of the World" with a stamp in commemoration by the U.S. Postal Service. The clone, called Pando (Latin for "I spread"), covers 106 acres with more than 40,000 individual trees—all originating from a single seed. Its exact age is not known, but it is believed to date from the end of the last ice age.

CHERRY (*Prunus* species)

Rose Family, Rosaceae

Native cherries in the *Prunus* genus were introduced in chapter 7, along with a description of their value to birds and other wildlife, propagation, and companion plants. Two cherry shrubs were described: choke cherry (*P. virginiana*) and sand cherry (*P. pumila*). Here two cherry trees will be covered: black cherry (*P. serotina*) and pin cherry (*P. pensylvanica*). These native trees are not to be confused with those in commercial orchards, which are mostly Eurasian species.

Both black cherry and pin cherry are found in forests and are considered gap or pioneer species, which readily take advantage of gaps or openings in the forest that occur when an old tree dies or when weather-related storms or fires create clearings. The trees can sprout wherever birds perch and feed on the fruits, dropping the seeds packed in fecal fertilizer to the ground, where the long-lasting seeds lie in the soil bank until opportunity arises.

Black cherries are a favorite fruit of Hermit Thrushes.
PHOTO BY RYAN BRADY

In spring these native cherry trees have lovely fragrant white flowers, one-half inch across. Black cherry showcases them in hanging clusters, while pin cherry's flowers occur in rounded clusters. Their barks are also distinctive and attractive. Mature black cherry trees have rough bark that looks like burnt cornflakes. Mature pin cherry's bark is shiny reddish brown and peels in horizontal strips. The beautiful berries of both trees entice both birds and humans. Pin cherry fruits are bright red, while black cherry's are dark red at first, turning black when fully ripe. Although the berries are devoured voraciously by birds, they can be quite bitter to the human palate (some black cherry trees are said to have sweeter-tasting berries). The berries of both trees are used to make jellies and wine and as a flavoring for rum and brandy ("cherry bounce").

A Closer Look

Tent caterpillars (*Malacosoma americanum*) and fall webworms (*Hyphantria cunea*) are among the most visible caterpillars found on native cherry trees, although they feed on other woody plants as well. In spring, tent caterpillars form large cobwebby tents in the crotches of cherry branches. The tents may contain up to 250 caterpillars. The caterpillars leave their tents several times a day to feed on cherry leaves. Later in the year, fall webworms build similar, but more gauzy, spiderweb-like nests surrounding several outer branches, making their nests even more visible than those of the tent caterpillar. Like tent caterpillars, they feed gregariously but stay within their tent to do so. Although these two caterpillars can sometimes defoliate the cherries, they rarely kill the healthy mature trees.

Cuckoos, handsome long-tailed birds, are especially fond of tent caterpillars and fall webworms—one of the few species that eat such spiny, hairy caterpillars. Cuckoo numbers tend to increase during outbreaks of these caterpillars as well as those of the spongy moth, formerly known as the gypsy moth. There are two species of American cuckoos, both rare. The Black-billed Cuckoo is on Audubon's watch list in sixteen states, and the western subspecies of the Yellow-billed Cuckoo is now listed as threatened under the Endangered Species Act. Cuckoos consume thousands of caterpillars per season; individual birds have been found with more than one hundred caterpillars in their stomachs. And they eat fast. One cuckoo ate thirty-six forest tent caterpillars in five minutes!

Yellow-billed Cuckoos feed voraciously on tent caterpillars and fall webworms.
PHOTO BY JEREMY MEYER, jeremymeyer.photography.

Check the cherry trees in your neighborhood for the tents of these two caterpillar species and you may be lucky to see cuckoos feasting on them. The Black-billed Cuckoo may shake and hammer a caterpillar on a branch or crush it in its bill before eating it. The yellow-billed rolls the caterpillars back and forth in its beaks before eating it. If you don't see the cuckoos, often hidden in the foliage, listen for their distinctive calls. The black-billed utters rhythmic series of *cucucu, cucucu* notes, usually grouped in threes or fours. The yellow-billed calls are single throaty notes sounding like *ka-ka-ka-ka-ka-kow-kow-kowlp-kowlp-kowlp-kowlp*, slowing toward the end. (Neither call is like an old-fashioned cuckoo clock, which mimics the sound of the European cuckoo.)

To ensure that these beautiful birds rebound in numbers and thrive once again, be sure to keep those caterpillar tents in your cherries. If neighbors complain, put up a sign highlighting their value to cuckoos, along with a picture of a cuckoo if possible.

Pin Cherry (*Prunus pensylvanica*)

Deciduous Tree
Size: Height 20–35 feet, width 10–15 feet
Color and Season: White flowers in spring; bright red to orangish foliage in autumn

Light Requirement: Sun
Soil: Moist to dry; well-drained sandy
Native Habitat: Forests, clearings, burned sites
Native to: CT CO DE IA IL MA MD ME MI MN MT ND NH NJ NY NC OH PA RI SD TN VA VT WI WV WY; native but rare in GA IN KY
USDA Hardiness Zones: 3–7

Pin cherry is sometimes called bird cherry because it attracts so many birds. Another common name is fire cherry due to its ability to quickly reforest a burned or cleared site; it forms thickets by sprouting from seeds in the forest soil that can remain viable for fifty to one hundred years. It has coppery-red bark and fiery fall color, the showiest of native cherry species. Pin cherry prefers cool climates, full sun, and moist, sandy soils. In these conditions, it is a hardy, adaptable tree although rather short-lived. Birds gorge on the abundant, shiny red berries, which are said to be especially valuable for Eastern Bluebirds. It's an excellent choice for gardens and naturalization.

Black Cherry (*Prunus serotina*)

Deciduous Tree
Size: Height 50–70 feet, width 20–30 feet
Color and Season: White flowers in late spring; yellow to orange foliage in autumn
Light Requirement: Sun
Soil: Moist to medium dry; fertile loam, clay loam
Native Habitat: Woodlands, savannas
Native to: AL AR AZ CT DE FL GA IA IL IN KS KY LA MA MD ME MI MN MO MS NC ND NE NH NJ NM NY OH OK PA RI SC TN TX VA VT WI WV
USDA Hardiness Zones: 3–9

Because of the wide native range of black cherry, there is a great deal of genetic diversity within the species. For example, seeds from trees in Wisconsin weigh more than those produced by black cherries in other states. As with many other plants, experts recommend purchasing trees that are native to your area, often referred to as local ecotype. Trees to the south or north of your location are

This Pileated Woodpecker is digging for insects in a black cherry tree.

PHOTO BY DAWN-MARIE STACCIA

unlikely to do well. Southern plants leaf out earlier than those to the north and are more likely to show winter damage. Northern plants will grow slower in the South, compared to the local ecotype trees native there. Locally sourced trees are very adaptable to many soil textures and moisture levels, as long as the area is not too wet or dry. Plant in full sun for the best growth. Black cherry can be used as a shade tree and for natural areas and restoration sites. Black cherry wood is one of the most valued in the country. Strong and hard, with beautiful reddish-brown color, the wood is used for cabinets and furniture.

OAK (*Quercus* species)

Beech Family, Fagaceae

> A yard without oaks is a yard meeting only a fraction of its life-support potential.
>
> —Douglas W. Tallamy, *The Nature of Oaks*

Oaks are champions among trees in their value to a variety of birds, hosting a banquet of insects and acorns. Planting an oak or group of oaks is arguably the best thing you can do to help birds. Among the largest and longest-lived trees, they are also the best trees to sequester carbon and help address climate change. See chapter 6 for information on the extraordinary value of all oaks

An Eastern Screech-Owl is well camouflaged in this oak tree.
PHOTO BY JEREMY MEYER, jeremymeyer.photography

for birds and wildlife, how to plant them, and other information. The savanna oaks featured in that chapter are black oak (*Quercus velutina*), bur oak (*Q. macrocarpa*), and Hill's oak (*Q. ellipsoidalis*). Here, we'll take a look at five oak species that occur primarily in woodlands and forests: white oak (*Q. alba*), swamp white oak (*Q. bicolor*), chinkapin oak (*Q. muehlenbergii*), pin oak (*Q. palustris*), and northern red oak (*Q. rubra*). Companion plants for these oaks are given in the descriptions here.

A Closer Look

Oaks have a lot of gall—literally. Galls are growths on the trees created by insects, and they come in an astonishing variety of shapes and sizes. Oaks host more types of gall insects than any other plants, and each oak species is host to various kinds of gall wasps. The galls can be found on nearly every part of an oak—leaves, twigs, branches, flowers, and roots. Most oak galls are formed by wasps that insert their eggs, along with plant growth regulators, into oak tissues. This results in a swelling of the oak tissue, surrounding the tiny egg with a safe chamber, where it grows, hatches, and feeds until it emerges as an adult wasp. These wasps do not sting and are harmless to humans. And mild infestations won't hurt your oak tree.

Many birds, including jays, nuthatches, titmice, and sapsuckers, forage on gall insects. Some birds even swallow the galls whole. Squirrels, chipmunks, and other mammals also feed on galls. In addition, there are wasps that specialize in parasitizing galls of other species, plus "hyperparasites" that parasitize the parasites. The world of oak galls is complex and plays a critical role in the food web.

Oak galls are relatively easy to find. Most common are oak apple galls, which look like little apples, one to two inches in diameter, and are found on many species of oak. Apple gall wasps of several species lay a single egg on a leaf, and the gall formed around it is green at first, becoming brown as the wasp matures. I've also found oak bullet galls on my white oaks. When fresh, these galls are a yellowish-red color. By mid to late summer, they become dark brown or red, looking like small bullets with smooth surfaces, and are found in clusters attached to twigs. These interesting galls are formed by the round bullet gall wasp (*Disholcapsis quercusglobulus*). How descriptive the last specific name—*quercus* (oak) and *globulus* (globule)! The adults within the galls emerge in fall and lay their eggs on oak twigs immediately. You can easily remove the galls for closer examination and check for the exit holes formed when the adult wasps emerge. Another white oak gall to search for is the lovely wool sower gall formed by a different wasp (*Callirhystis seminator*). It's a fuzzy white ball spotted with red or dark brown spots up to two inches in diameter. I'm keeping a lookout for this beauty, as are other gall lovers I know. See more on galls in the description of red oaks.

Historical note: Galls were used to make the ink used in writing the Declaration of Independence and the U.S. Constitution. It consisted of tannic acid from oak galls mixed with iron from nails or iron scraps, with a binder like gum arabic.

White Oak (*Quercus alba*)

Deciduous Tree
Size: Height 75–100 feet, width 75–100 feet
Color and Season: Yellow-green catkins in spring; burgundy foliage in autumn
Light Requirement: Sun to partial shade
Soil: Prefers moist, tolerates dry; adaptable to a variety of soils
Native Habitat: Woodlands, savannas

Native to: AL AR CT DE FL GA IA IL IN KS KY LA MD ME MI MN MO MS NC NH NJ NY OH OK PA RI SC TN TX VA VT WI WV; rare in NE
USDA Hardiness Zones: 3–9

White oak is a beautiful picturesque tree, and we're fortunate to have several in our yard. I'm not sure who loves them more—we or the birds. Nearly every year, robins nest in the boughs of a white oak just outside one of our windows. No need for a live cam. We can watch the progression of nest building and egg incubation—at least until the oak foliage gives them more privacy. Best of all, the white oak attracts many migrants to our yard in spring when the tree hosts an abundance of caterpillars for the famished travelers. This oak alone hosts 436 caterpillar species! It's especially a thrill when warblers arrive. Called the "butterflies of the bird world" for their colorful plumage, warblers may visit for only a few days before continuing northward to their breeding areas in northern Wisconsin and Canada. A few years ago, Connecticut and Mourning Warblers, very rare warblers, were among those that visited in late May, and I was able to watch these secretive birds only a few feet away as they busied themselves foraging in the oak leaf litter and low vegetation. Another year, I simply sat on a lawn chair in my driveway beside an ancient white oak and enjoyed seeing and hearing forty species of birds, including many warblers, within a few hours. To be sure, such amazing days are rare. Some years the migrants simply fly over to their northerly destination, or the weather is not to their liking. Or like the last few years, the extreme weather in the South reduced their numbers

A Rose-breasted Grosbeak forages for insects in an oak tree.
PHOTO BY JACK BARTHOLMAI

and migrants were far fewer. To ensure such magic days occur in the years to come, oaks are a must in the yard of any bird lover. And climate change, which is causing the severe weather events harming both birds and us humans, needs to be addressed. Companion plants that have been documented growing beneath white oaks include lovely herbaceous plants such as round-leaved hepatica (*Hepatica americana*), Canada mayflower (*Maianthemum canadense*) and false Soloman's seal (*M. racemosum*).

Swamp White Oak (*Quercus bicolor*)

Deciduous Tree
Size: Height 60–80 feet, width 65 feet
Color and Season: Yellow-green catkins in spring; golden-brown foliage in autumn
Light Requirement: Sun, part sun
Soil: Moist to wet; prefers rich, slightly acidic soils; tolerant of heavy clay, slightly alkaline, even compacted soils
Native Habitat: Swampy and floodplain woodlands
Native to: CT DE IA IL IN KY MA MD MI MN MO NH NJ NY OH PA RI TN VA VT WI WV; native but rare in AL ME NC NE SC
USDA Hardiness Zones: 3–8

Swamp white oak is one of the best oaks for wet soils. Its shallow fibrous roots, typical of wetland trees, make it easy to transplant. The Latin species name *bicolor* means two colors and refers to its leaves, which are glossy green above and fuzzy white below. Swamp white oak prefers slightly acidic conditions, but it is adaptable to medium-dry soils that are neutral or slightly alkaline. In highly alkaline soils, a better choice is the naturally occurring hybrid swamp x bur oak, which is more alkaline adaptable. Both swamp white oak and the hybrid have large round crowns and are excellent, long-lived shade trees for large yards, parks, and golf courses. Companion plants found growing beneath this oak include several hummingbird plants highlighted in chapter 4: orange jewelweed (*Impatiens capensis*) and Michigan lily (*Lilium michiganense*). Another species often occurring with this oak is the beautiful blue flag iris (*Iris viriginica shrevei*), which also attracts hummingbirds.

Chinkapin or Chinquapin Oak (*Quercus muehlenbergii*)

Deciduous Tree
Size: Height 40–60 feet, width 40–60 feet
Color and Season: Yellow-green catkins; yellow-brown foliage in autumn
Light Requirement: Sun
Soil: Dry to medium dry; requires alkaline; adaptable to most soils except wet
Native Habitat: Woodlands, savannas
Native to: AL AR CT DE FL GA IA IL IN KS KY LA MD ME MI MO MS NE NH NJ NM NY OH OK PA RI SC TN TX VA WV; native but rare in MA MN NC NJ VT WI
USDA Hardiness Zones: 5–7

Chinkapin oak has one of the widest ranges of any oak, but it is seldom a dominant tree and is occasional to rare in some states. Its common name refers to the fact that its leaves are not lobed, like most oaks, but are toothed like those of the American chestnut (*Castanea dentata*), which was also known as chinkapin. The Latin species name honors Henry Ernst Muehlenberg (1753–1815), a Pennsylvania botanist. This attractive oak is extremely hardy and thrives in a wide range of sunny habitats, from woodlands to barrens. It is adaptable to most soils, as long as they are not too wet or acidic. Its nuts are the sweetest among oaks and were prized by American Indians and are favored by wildlife as well. The tree is also an important host for the caterpillars of the gray streak butterfly. Companion plants commonly found with this oak are these showy herbaceous species: wild geranium (*Geranium maculatum*), elm-leaved goldenrod (*Solidago ulmifolia*), and Short's aster (*Symphyotricum shortii*).

Pin Oak (*Quercus palustris*)

Deciduous Tree
Size: Height 50–80 feet, width 25–35 feet
Color and Season: Yellow-green catkins in spring; red to burgundy foliage in autumn
Light Requirement: Sun, part sun
Soil: Wet; tolerates average moisture; somewhat acidic sand, clay, loam
Native Habitat: Floodplains, swamp forests

Native to: AL AR CT DE FL GA IA IL IN KS KY MA MD MI MO NJ NY OH OK PA RI SC TN VA WV; native but rare in MS NC WI; extirpated in GA
USDA Hardiness Zones: 4–8

Like many wetland trees, this oak has shallow fibrous roots, not a taproot as many upland oak species have, making it is easier to transplant. It has a graceful appearance, with a strongly pyramidal shape, and is one of the most widely planted and popular oaks. Named for its many short side twigs or pin-like spurs, pin oak loves wet feet and tolerates average moist soils but is sometimes misplaced in drier conditions where it does poorly. In our former home, we planted a pin oak that did very well in our clay soil. Drainage from our sump pump helped maintain moist conditions. This species grows faster than most oaks, and a healthy tree can live 100 to 150 years. Its small acorns are a perfect size for many nut-eating birds. Companion plants commonly found with pin oak grown in moist soils include three species highlighted in previous chapters: wild bergamot (*Monarda fistulosa*), false sunflower (*Heliopsis helianthoides*), and cup plant (*Silphium perfoliatum*).

Northern Red Oak (*Quercus rubra*)

Deciduous Tree
Size: Height 75 to 100 feet, width 60–75 feet
Color and Season: Green catkins in spring; rich brownish-red foliage in autumn
Light Requirement: Sun, partial shade
Soil: Moist, medium dry; prefers loam; adapts to soil containing clay, sand, gravel
Native Habitat: Moist woodlands, ravines, north and east slopes
Native to: AL AR CT DE GA IA IL IN KS KY MA MD ME MI MN MO MS NC NE NH NJ NY OH OK PA RI SC TN VT VA WI WV; native but rare in LA
USDA Hardiness Zones: 4–8

Red oaks have been called monarchs of mature deciduous forests in the Midwest and Northeast, preferring rich, moist forest soils. Well named, this oak has a beautiful reddish heartwood, as well as rich mahogany leaves in fall. The bark of older trees is quite distinctive with wide, flat furrows that look as if a

The red oak has rich mahogany leaves in fall.
PHOTO BY MARIETTE NOWAK

child finger-painted down the trunk—an easy identification feature. This species is one of the fastest-growing oaks and is relatively easy to transplant due to a negligible taproot. It is also hardy in urban conditions, making it a great selection for nearly any yard with fairly moist soils and enough room to accommodate its width. Plant as a specimen tree or in groups to create a mini-woodland corner in your yard. Companion plants that do well with red oak include wild geranium (*Geranium maculatum*), woodland phlox (*Phlox divaricata*), and elm-leaved goldenrod (*Solidago ulmifolia*). The red oaks offer birds a nutritious menu of treats all year long. As you'll see here, autumn and winter specialties include gall insects, inchworms, and probably many more entrees we've yet to discover.

A few years ago in November, I noticed that many of the goldfinches that usually devour seeds at my feeders and feast on the seed heads in my prairie plants were also foraging busily among the leaves of a red oak. What were they doing? What had they found? Some leaves were completely brown and crisp, while others still retained some green chlorophyll. But all were loaded with many small, round bumps, as many as nine or more per leaf. Wisconsin entomologist Phil Pellitteri identified them as the galls created by cynipid gall wasps. These wasps induced my oak to form a small, safe chamber—a gall—around the eggs they laid on the leaves. At least ten species of cynipid gall wasps can be found on red oaks. The finches were probing the galls and feasting on the protein-packed insects within. Since then, I watch for these galls on my oaks and have the pleasure of knowing that my finches have a well-balanced diet of both seeds and bugs over the winter months.

In winter, the Golden-crowned Kinglet survives on tiny caterpillars it finds on trees.
PHOTO BY JEREMY MEYER, jeremymeyer.photography

Besides nuts and galls, are there any other winter foods for birds? What are Brown Creepers and other birds doing creeping up and down the trunks of my oaks and other trees in the midst of subzero winter days? Nothing, seemingly, is there for them. Are they simply exercising to keep warm? Bernd Heinrich, a noted naturalist and author in Maine, was intrigued by the mystery. To find out, he dissected the crops of Golden-crowned Kinglets, tiny birds barely larger than hummingbirds, that had crashed into a window and died. He discovered that they were far from starving—their tiny crops were stuffed with caterpillars. Many moth species winter in the caterpillar stage. Inchworms—caterpillars of the moth *Lytrosis unitaria*, for example—winter in bark crevices or lie stock still on tree branches, protecting themselves with glycerin, the same ingredient in car antifreeze. For birds like kinglets and creepers that feed only on insects, these caterpillars are a critical part of their winter diet. But other "feeder" birds, including chickadees, nuthatches, titmice, and some woodpeckers, seek out these insects as well, supplementing seeds with insects.

Appendix

Top Wetland and Rain Garden Plants for Birds

A number of the recommended plants for birds featured in this book are excellent choices for wet areas in various habitats, including hummingbird gardens, prairies, shrublands, and woodlands. They will also do well in rain gardens and various wetlands, such as swales and the shores of ponds, rivers, and lakes.

Herbaceous Species

swamp or fen thistle (*Cirsium muticum*)
jewelweed (*Impatiens capensis*)
marsh gay feather (*Liatris spicata*)

The beloved Red-winged Blackbird is a harbinger of spring, announcing his arrival with a lively gurgling konk-la-ree.
PHOTO BY JACK BARTHOLMAI

Michigan lily (*Lilium michiganense*)
cardinal flower (*Lobelia cardinalis*)
pussy willow (*Salix discolor*)
cup plant (*Silphium perfoliatum*)
New England aster (*Symphyotrichum novae-angliae*)

Shrubs

red osier dogwood (*Cornus sericea*)
blue-fruited or swamp dogwood (*Cornus obliqua*)
ninebark (*Physocarpus opulifolius*)
beaked willow or Bebb's willow (*Salix bebbiana*)
common elderberry (*Sambucus nigra* subsp. *canadensis*)
highbush blueberry (*Vaccinium corymbosum*)

Trees

silver maple (*Acer saccharinum*)
river birch (*Betula nigra*)
eastern cottonwood (*Populus deltoides*)
swamp white oak (*Quercus bicolor*)

Bibliography

Benyus, Janine M. *Northwoods Wildlife*. NorthWord Press, 1989.

Burghardt, Karin T., Douglas W. Tallamy, Christopher Philips, and Kimberley J. Shropshire. "Non-Native Plants Reduce Abundance, Richness, and Host Specialization in Lepidopteran Communities." *Ecosphere* 1, no. 5 (2010): 1–22.

Chalker-Scott, Linda. "Wood Chip Mulch: Landscape Boon or Bane?" *MasterGardener*, Summer 2007. https://s3.wp.wsu.edu/uploads/sites/403/2015/03/wood-chips.pdf.

Cullina, William. *Native Trees, Shrubs, and Vines: A Guide to Using, Growing, and Propagating North American Woody Plants*. Houghton Mifflin, 2002.

Cullina, William. *The New England Wildflower Society Guide to Growing and Propagating Wildflowers of the United States and Canada*. Houghton Mifflin, 2002.

Diboll, Neil, and Hilary Cox. *The Gardener's Guide to Prairie Plants*. University of Chicago Press, 2023.

Eckberg, James, Eric Lee-Mäder, Jennifer Hopwood, Sarah Foltz Jordan, and Brianna Borders. *Native Thistles: A Conservation Practitioner's Guide*. Xerces Society for Invertebrate Conservation, 2017.

Griscom, Bronson W., et al. "Natural Climate Solutions." *PNAS* 114, no. 44 (2017): 11645–50.

Heinrich, B., and R. Bell. "Winter Food of a Small Insectivorous Bird, the Golden-Crowned Kinglet." *Wilson Bulletin* 107, no. 3 (1995): 558–61.

Hicks, Damien M., et al. "Food for Pollinators: Quantifying the Nectar and Pollen Resources of Urban Flower Meadows." *PLOS One* 11, no. 6 (2016): e0158117.

Hightshoe, Gary L. *Native Trees, Shrubs, and Vines for Urban and Rural America: A Planting Design Manual for Urban Designers*. John Wiley and Sons, 1988.

Johnson, Carter, and Curtis S. Adkisson. "Airlifting the Oaks." *Natural History*, October 1986.

Kaufman, Kenn. *Lives of North American Birds*. Houghton Mifflin, 1996.

Kennedy, Ashley C. "Examining Breeding Bird Diets to Improve Avian Conservation Efforts." PhD diss., University of Delaware, 2019.

Kim, Ashley Y., David T. Rankin, and Erin E. Wilson. "What Is That Smell? Hummingbirds Avoid Foraging on Resources with Defensive Insect Compounds." *Behavioral Ecology and Sociobiology* 75, no. 9 (2021): 132.

Leopold, Aldo. *A Sand County Almanac*. Oxford University Press, 1996.

Loss, Scott, Tom Loss, Sara S. Loss, and Peter Marra. "Bird–Building Collisions in the United States: Estimates of Annual Mortality and Species Vulnerability." *The Condor* 116, no. 1 (2014): 8–23.

MacArthur, Robert H. "Population Ecology of Some Warblers of Northeastern Coniferous Forests." *Ecology* 39, no. 4 (1958): 599–619.

Marra, Peter P., and Chris Santella. *Cat Wars: The Devastating Consequences of a Cuddly Killer*. Princeton University Press, 2016.

Martin, A. C., Herbert S. Zim, and Arnold L. Nelson. *American Wildlife and Plants: A Guide to Wildlife Food and Habits*. Dover, 1951.

McGuire, Andrew. "Comparing Effects of Herbicides, Fertilizers, and Tillage on the Soil." Washington State University Center for Sustaining Agriculture and Natural Resources. 2016. https://csanr.wsu.edu/comparing-effects-on-soil/.

Mineau, Pierre, and Cynthia Palmer. "The Impact of the Nation's Most Widely Used Insecticides on Birds." *American Bird Conservancy*, March 2013. https://dariuszzdziebk.wpenginepowered.com/wp-content/uploads/2015/05/Neonic_FINAL.pdf.

Narango, Desiree L., Douglas W. Tallamy, and Peter P. Marra. "Native Plants Improve Breeding and Foraging Habitat for an Insectivorous Bird." *Biological Conservation* 213 (September 2017): 42–50.

Narango, Desiree L., Douglas W. Tallamy, and Peter P. Marra. "Nonnative Plants Reduce Population Growth of an Insectivorous Bird." *Proceedings of the National Academy of Sciences* 115 (2018): 11549–54.

North American Bird Conservation Initiative. "State of the Birds Report: United States of America, 2022." http://www.stateofthebirds.org/2022/.

Nowak, Mariette. *Birdscaping in the Midwest*. University of Wisconsin Press, 2012.

Peattie, Donald Culross. *A Natural History of Trees*. Houghton Mifflin, 1950.

Peterson, Lee Allen. *Edible Wild Plants*. Houghton Mifflin, 1977.

Pörtner, Hans-Otto, et al. "IPBES-IPCC Co-Sponsored Workshop Report on Biodiversity and Climate Change." IPBES and IPCC, June 24, 2021. DOI:10.5281/zenodo.4782538.

The Prairie Enthusiasts. *The Prairie Promotor* 36, no. 2 (Summer 2023). https://theprairieenthusiasts.org/wp-content/uploads/2023/07/FINAL.WEB-ThePrairieEnthusiasts_ThePrairiePromoter_Summer2023.7.pdf.

Richard, Melissa, Douglas W. Tallamy, and Adam B. Mitchell. "Introduced Plants Reduce Species Interactions." *Biological Invasions* 21, no. 3 (2018): 983–92.

Rico-Guevara, Alejandro, Tai-hsi Fan, and Margaret A. Rubega. "Hummingbird Tongues Are Elastic Micropumps." *Proceedings of the Royal Society of Biological Sciences* 282, no. 1813 (2015).

Roebling, A. D., et al. "Rabies Prevention and Management of Cats in the Context of Trap–Neuter–Vaccinate–Release Programmes." *Zoonoses and Public Health* 61, no. 4 (2014): 290–96.

Russo, Ronald A. *Plant Galls of the Western United States*. Princeton University Press, 2021.

Salmón, Enrique. *Iwígara, American Indian Ethnobotanical Traditions and Science*. Timber Press, 2020.

Scott, Thomas. "Jays Plant Acorns." *Oaks 'n Folks* 5, no. 2 (November 1990). https://oaks.cnr.berkeley.edu/jays-plant-acorns/.

Shaw, Daniel, and Rusty Schmidt. *Plants for Stormwater Design: Species Selection for the Upper Midwest*. Minnesota Pollution Control Agency, 2003.

Tallamy, Douglas W. *Bringing Nature Home*. Timber Press, 2009.

Tallamy, Douglas W. *The Nature of Oaks*. Timber Press, 2021.

Tallamy, Douglas W. *Nature's Best Hope*. Timber Press, 2019

Tallamy, D. W., and K. J. Shropshire. "Ranking Lepidopteran Use of Native Versus Introduced Plants." *Conservation Biology* 23, no. 4 (2009): 941–47.

Wilhelm, Gerould, and Laura Rericha. *Flora of the Chicago Region: A Floristic and Ecological Synthesis*. Indiana Academy of Science, 2017.

Websites Consulted

The Biota of North America Program
http://www.bonap.org
Birds of the World
https://birdsoftheworld.org/bow/home
BugGuide
http://bugguide.net
Cornell Lab of Ornithology, All About Birds
https://www.allaboutbirds.org/guide/search
Flora of North America
http://www.eFloras.org
Flora of Wisconsin
https://wisflora.herbarium.wisc.edu/index.php
Homegrown National Park
https://homegrownnationalpark.org
Illinois Wildflowers
http://www.illinoiswildflowers.info
In Defense of Plants
https://www.indefenseofplants.com
Integrated Taxonomic Information System
http://www.itis.gov
Lady Bird Johnson Wildflower Center, Native Plant Base
http://wildflower.org/plants
Michigan Flora
http://www.michiganflora.net
Minnesota Wildflowers
http://www.minnesotawildflowers.info
Missouri Botanical Gardens, Plant Finder
http://www.missouribotanicalgarden.org/plantfinder

The Morton Arboretum
https://mortonarb.org
National Wildlife Federation, Native Plant Finder
https://www.nwf.org/nativeplantfinder/
National Wildlife Federation, Native Plant Habitats
https://www.nwf.org/Native-Plant-Habitats/Plant-Native
Perfect Earth Project
https://perfectearthproject.org
U.S. Department of Agriculture, Celebrating Wildflowers
https://www.fs.usda.gov/managing-land/wildflowers
U.S. Environmental Protection Agency, Ecoregions
https://www.epa.gov/eco-research/ecoregions
The Young Forest Project
https://youngforest.org

Index

Note: Page numbers in italics indicate illustrations.